The Institute of Biology's
Studies in Biology no. 107

Pollen and Allergy

R. Bruce Knox
Professor of Botany
University of Melbourne

© R. Bruce Knox, 1979

First Published 1979 by Edward Arnold (Publishers) Ltd, London
First Published in the USA in 1979 by
University Park Press
233 East Redwood Street
Baltimore, Maryland 21202

Library of Congress Cataloging in Publication Data

Knox, R. Bruce.
  Pollen and allergy.

  (The Institute of Biology's studies in biology;
no. 107)
  Bibliography: p.
  1. Pollen. 2. Allergens. I. Title.
II. Series: Institute of Biology. Studies in
biology; no. 107.
QK658.K63   1978   582'.0463   78-11594

ISBN 0-8391-0257-7

All rights reserved. No part of this publication may be reproduced, stored in a retrieval system, or transmitted in any form or by any means, electronic, photocopying, recording or otherwise, without the prior permission of Edward Arnold (Publishers) Ltd.

Printed in Great Britain

General Preface to the Series

Because it is no longer possible for one textbook to cover the whole field of biology while remaining sufficiently up to date the Institute of Biology has sponsored this series so that teachers and students can learn about significant developments. The enthusiastic acceptance of 'Studies in Biology' shows that the books are providing authoritative views of biological topics.

The features of the series include the attention given to methods, the selected list of books for further reading and, wherever possible, suggestions for practical work.

Readers' comments will be welcomed by the Education Officer of the Institute.

1979

The Institute of Biology,
41 Queen's Gate,
London SW7 5HU

Preface

Pollen grains, the structures housing the male gametes of plants, have been associated with man throughout his history. This little book provides an introduction to the life of the pollen grain and, in particular, deals with its dual role in plant fertilization and human allergic disease.

In the last decade, our understanding of the function of the intricately-sculptured outer walls of pollen grains has greatly advanced. They contain informational molecules that act as recognition factors. These determine, on the one hand, acceptance of an appropriate mate by the female stigma for seed-setting, and on the other hand, the onset of seasonal asthma and hay fever in susceptible humans. The mechanisms behind these responses are presented together with an account of pollen dispersal by air currents in city atmospheres and in the pollen load of honey bees.

I thank Professor J. Heslop-Harrison, F.R.S., for first stimulating my interest in pollen, my colleagues, especially Dr A. Clarke and Mrs S. Ducker, for helpful appraisal of the manuscript, Anne Pottage for assistance with Fig. 2–5 and Marg Robertson for secretarial assistance.

Melbourne, 1978

R. B. K.

Contents

General Preface to the Series   iii

Preface   iii

1 **The Pollen Grain**   1
   1.1 Pollen in history   1.2 Form of pollen grains   1.3 Pollen-wall structure   1.4 Pollen analysis and the history of vegetation

2 **Formation of Pollen**   11
   2.1 Meiosis and its consequences   2.2 The puzzle of wall pattern   2.3 Pollen development   2.4 Cell biology of the pollen grain

3 **Dispersal of Pollen**   22
   3.1 Adaptations for wind dispersal   3.2 Pollen transport by insects and other invertebrates   3.3 Pollination by birds, bats and mammals   3.4 Underwater pollination

4 **Pollen and Fertilization**   33
   4.1 The two domains of the pollen wall   4.2 Male–female recognition   4.3 Control of pollination

5 **Aerobiology of Pollen**   39
   5.1 Pollen transport in the atmosphere   5.2 The pollen calendar   5.3 Circadian rhythms of pollen emission

6 **Pollen and Man**   49
   6.1 Role of pollen in allergic disease   6.2 Allergens in pollen   6.3 Role of pollen allergens in nature

7 **Experiments with Pollen**   58
   7.1 Making a reference collection of pollen types   7.2 Pollen viability   7.3 Demonstration of pollen-wall proteins   7.4 Demonstration of pollen grains in honey   7.5 Pollen germination and demonstration of callose

Further Reading   60

1 The Pollen Grain

The pollen grain is the specialized structure which houses the sperm or male gametes of flowering plants. The word pollen was introduced by the great Swedish botanist, Linnaeus, in 1760 and is derived from the Latin root, fine flour, referring to its dry granular nature. It comprises two or three cells combined as a unit, and typically contains approximately 20% protein, 37% carbohydrate, 4% lipid and 3% mineral composition.

1.1 Pollen in history

Man has been aware of pollen throughout recorded history. The earliest records are contained in prehistoric artworks, stone carvings and bricks from the palaces of the Assyrian kings in the Hammurabi period (800 B.C.) and demonstrate early recognition of the sexes in plants. They illustrate mythological giant winged creatures apparently pollinating the female inflorescences of the date palm by shaking the male inflorescences over them. Herodotus, during his travels in Assyria, recognized the dioecious nature of the date palm, but this was later dismissed by Aristotle on the grounds that plants were not motile and had no need for separation of the sexes. His pupil, Theophrastus, had no such delusions, stating that 'the fruit of the female date palm does not perfect itself unless the blossom of the male with its dust is shaken over it'. He concluded that this process of fertilization was not unique to the date, and was likely to occur in all plants.

Theophrastus noted the parallels between this and another ancient practice, the caprification of figs. The cultivated fig, *Ficus carica*, has hollow fruits or syconia containing female flowers. For fruit-setting, branches of the wild goat fig (caprifig) are tied onto the trees at flowering. The fruits of the goat fig contain fertile anthers at the top and sterile female flowers below which are hosts for the gall wasp larvae. Pollination is carried out by the gall wasps which emerge from the female flowers within the goat fig fruits and, in order to escape, have to push through the dehiscing anthers which surround the exit pore. They carry the pollen on their body surface to the receptive female fruits of the cultivated figs. Theophrastus' report indicates some understanding of the role of insects in pollination.

Man's interest in pollen at that time, and in the succeeding Dark Ages, was largely for its pharmacological properties. In the sixteenth century, Turner, in his herbal, repeated the Greek Dioscorides' note that the '*floure*' (pollen) that is found in the centre of a rose 'is good against the reume or flowing of the gummes'. Little progress in understanding the real nature of pollen was made until the seventeenth century with the advent of modern scientific inquiry. The sexual nature of the flower was foreshadowed by the English plant anatomist, Nehemiah Grew. Grew aptly described dehiscence of the anther: 'At these clefts, it is that they disburse their powders; which as they start out, and stand between the two lips of each cleft have some resemblance to the common sculpture of a pomegranate with the seeds looking out at the cleft of its rind.' Those who have examined a ripe pomegranate will appreciate his analogy. The presence of furrows on the surface of lily pollen was noted by the Italian biologist, Malpighi, who considered pollen as a mere secretion prior to maturation of the ovary.

The breakthrough in recognizing pollen as the male element in plant life was made by Rudolf Camerarius, a professor at Tubingen. He experimented with anther manipulation and emasculation in species such as dog's mercury, *Mercurialis*, and caster bean, *Ricinus*, and concluded in his book published in 1694: 'in plants, no production of seeds takes place unless the anthers have prepared the young plant in the seed. It is thus justifiable to regard the anthers as male, while the ovary with its style represents the female part . . . the stamens are the male sexual organs in which that powder which is the most subtle part of the plant is secreted and collected.'

This pioneering work inspired others to carry out experiments on pollination. James Logan, governor of Pennsylvania in 1739, set up trials of corn, *Zea mays*, demonstrating that pollen from the tassels travelled in air currents to the cobs. The role of insects in pollination was demonstrated by Joseph Koelreuter, director of the grand-ducal botanic garden at Carlsruhe, in experiments published between 1761 and 1766. He showed that insects may be attracted to flowers by their nectar, which bees used to make honey. Later, Christian Sprengel, rector of Spandau, became so involved in pollination experiments that he was accused of neglecting his parish. He established the role of floral nectaries in producing nectar as an attractant for insect pollination, and noted that nectar-guides, coloured markings on the flower petals, are placed to indicate the sites of nectaries. He made remarkable observations on flowers which imprison and even destroy the insects that pollinate them.

Despite these advances in our understanding of pollination, little progress was made in coming to grips with the mechanism of fertilization. The concepts were rather vague, perhaps because of the limitations in the scientific instruments of the day. Pollen was said to burst on the stigma, releasing its fertilizing granules which found their way through stylar channels to the ovary (Needham, 1740). Koelreuter was more perceptive. He reported that the oily fluid secreted from the surface of pollen mixed with stigma secretion to form a new substance which flowed through the style to the ovule where he believed fertilization occurred.

All these concepts were to change at the turn of the nineteenth century, with the availability of new and more powerful microscopes, which gave both higher magnification (up to 500 times) and higher resolution. The French artist and botanist, Turpin and an Italian microscopist, Amici, discovered independently in the 1820s the existence of the pollen tube. Amici followed this work with the discovery of the nutritional dependence of the pollen tube on the pistil, in 1846. Its role as the carrier of the sperm to the ovule was demonstrated later by the French botanist Brogniart.

At this time, great progress was being made in elucidating the structure of the pollen grain. Koelreuter, in the mid-eighteenth century, had considered that the pollen grain had a cellular core covered by two coats, the outer being tough and elastic. This was confirmed by Mirbel, who noted the presence of germinal apertures. Fine details of pollen wall structure were obtained by Hugo von Mohl, who introduced sectioning procedures, which superseded the former tearing or macerating methods. He developed a classification of pollen based on the geometrical configuration of the apertures. Working at St Petersburg in 1833, Julius Fritsche produced the first natural classification of pollen based on both morphological and chemical studies of the pollen walls. He named the outer patterned wall *exine*, and the inner smooth wall *intine*, and demonstrated their differential solubility in sulphuric acid. Later, Fischer showed a progression in thickening and complexity of exine structure, and increased prominence and number of germinal apertures in pollen of more highly evolved species.

In the final quarter of the nineteenth century, the last hurdle to understanding the nature of pollen was overcome: the existence of the sperm cells and their origin through the alternation of generations was established. As early as 1842, the apothecary Wilhelm Hofmeister had demonstrated spermatozoa in several lower plants, even before their discovery in animal systems; but it was not until the work of Celakovsky in 1874 and, later, Strasburger that proof of the alternation of generations in flowering plants was obtained. Controversy surrounded the cellular nature of the sperm, which was not resolved until 1965, with the advent of the transmission electron microscope. The remaining landmarks of the twentieth century form the basis for this book and will be considered in the chapters that follow.

1.2 Form of pollen grains

Pollen occurs in a number of shapes, mostly variations of spheres, and in sizes, ranging up to about 300 μm (Figs 1–2, 1–3 and 1–4). Their geometry is determined to some extent by the number and position of the germinal apertures: round to spherical grains having 0–many apertures, long grains having two apertures, triangular (3-sided) grains having three apertures and quadrate (4-sided) grains having four apertures.

The apertures are a major feature of grain morphology. Apertures may be long furrows, pores or a combination of the two structures (Fig. 1–1).

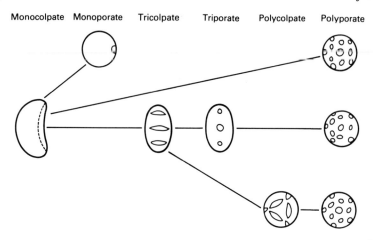

Fig. 1–1 Evolution of pollen shape and type in angiosperms. (Adapted from TAKHTAJAN, A. (1959). *Die Evolution der Angiospermen*. Jena.)

Pores are circular structures, while furrows are defined as having their long axes more than twice their width. Pollen grains show a distinct polarity in organization, with the furrows running longitudinally down through the equator of the grain from the poles at either end. Thus the polar view (from the top or bottom of the grain) gives a distinctly different impression of morphology from the equatorial view (from the sides). In polar view, as determined by the convergence of the furrows towards the pole, the equator is essentially the outline of the grain.

Monocotyledons generally have pollens with a single aperture, usually a broad furrow termed a colpus. In dicotyledons, three, four or five

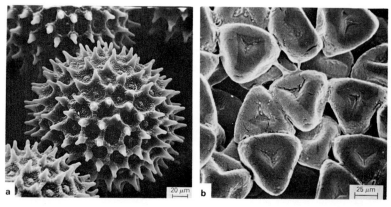

Fig. 1–2 Pollen grain shapes. (a) Morning glory, *Ipomaea*, pollen grain, showing germinal apertures surrounded by spines. (b) *Eucalyptus* pollen grain showing germinal apertures at corners of triangles. Grains are covered with pollenkitt.

apertures are commonly found, and may be pores or a combination of pores with furrows, described as *colporate* (Figs 1–2, 1–3 and 1–4). These are usually radially symmetrical about the equator. Other dicotyledons have many apertures, and in these cases, they are usually pores. Pores are often covered by a lid or operculum of sporopollenin, which is shed at germination. The scheme suggested by A. Takhtajan in 1959 for the evolution of pollen aperture types is shown in Fig. 1–1.

Another dimension of form appears when individual pollen grains adhere together to form polyads or pollen masses of various kinds. Tetrad or polyad grains result when the tetrad of microspores remain in contact after the dissolution of the callose special wall that surrounds them at the end of meiosis. Tetrads of grains are typical of certain

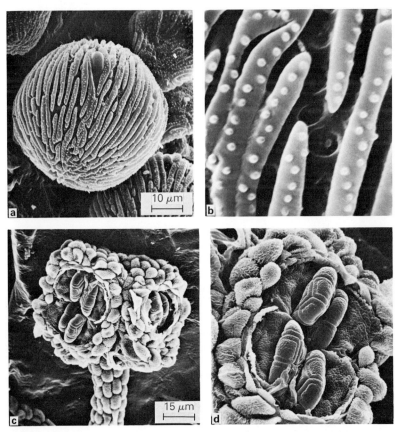

Fig. 1–3 Form of pollen grains. (a–b) Trumpet lily, *Datura*, pollen grain. (a) Shows striate exine surface patterning. One of the three germinal apertures is visible. (b) Detail of exine surface showing porous nexine. (c–d) Anther of *Acacia subulata*, an Australian wattle, showing dehiscence of polyads, each of 16 grains.

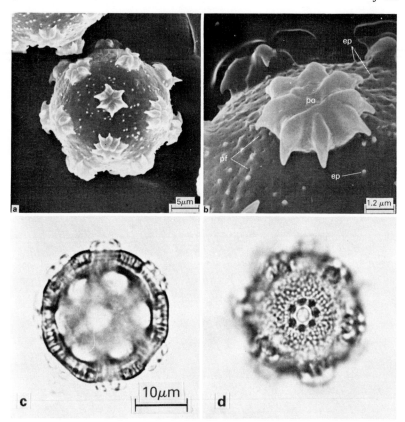

Fig. 1–4 Unusual sculptured pollen grains of *Pupalia lappacea*, Amaranthaceae. The spherical grains have ornamented germinal apertures (po). The exine is ornamented with small spines (ep) and perforated by micropores (pf). (c) and (d) show a surface and equatorial view of the same grains in the light microscope using phase contrast optics after they have been cleaned by acid treatment (acetolysis). (From ZANDONELLA, P. and LECOCQ, M. (1977). *Pollen et Spores*, **19**, 119.)

families, for example the heath families, Epacridaceae, Ericaceae, the sundews, Droseraceae, and rushes, Juncaceae, and of certain genera within the Australian families Goodeniaceae and the acacias, Mimosaceae. It is not known why the grains adhere, but in tetrads of *Lechenaultia* (Goodeniaceae) the rods of the exine connect the grains, suggesting that close contact may be established early in development. Monads, dyads, triads and nullads can occur when one, two, three or all the grains are sterile. In the mimosa, *Acacia*, 4, 8, 16, 32 or 64 grains regularly associate in polyads (Fig. 1–3), depending on the number of meiotic tetrads held within segments of the anther.

Pollen grains associate in masses in orchids (Orchidaceae) and milkworts (Asclepiadaceae). In orchids, an array of pollen associations is found from free individual grains in apostasioid orchids, adhesive masses in the lady's slippers, cypripediloid orchids, and tetrads loosely united by elastic viscin threads of tapetal origin in the orchidoid and neottioid orchids (Fig. 1–5). The orchidoids possess pollen sacs organized into many interconnected packets or massulae. In the neottioids, all the pollen within a sac is associated together into large structures called pollinia which can be easily separated into mealy or powdery masses. In the higher orchids, such as the Ophrydeae, pollinia may be hard masses, each of which is a separate dispersal unit.

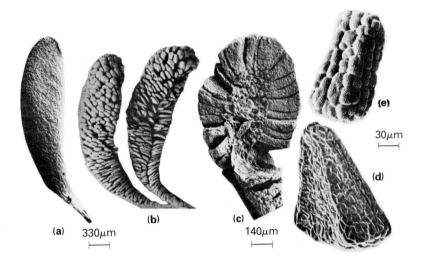

Fig. 1–5 Structure of pollinia and massulae of orchids as seen by scanning electron microscopy in (a) *Spiranthes*, (b) *Haemaria*, and (c) *Ophrys* which all have pollen sacs arranged into discrete pollinia, (d) *Orchis* and (e) *Gymnadenia* where the pollen sacs are aggregated into massulae. (From SCHILL, R. and PFEIFFER, W. (1977). *Pollen et Spores*, **19**, 5–118.)

1.3 Pollen-wall structure

The pattern of the exine is determined by the kind of structural elements of which it is composed. There are two basic types of exine structure (Fig. 1–6). In *pilate* types, the rods are surmounted by a prominent knob which may be fused together to form intricate patterns, for example, the net-like exine of lily and crucifers. In *tectate* types, the rods are covered over by a roof which is often ornamented by spines, knobs and other features (Fig. 1–5). The outer covering is perforated by micropores, often concentrated around the base of spines giving access

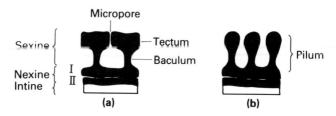

Fig. 1-6 Diagrams of the stratification of pollen walls; (a) shows a wall with a tectate exine and (b) with a pilate exine. (Modified from HESLOP-HARRISON, J. (1968). *Science*, N.Y., **161**, 230.

to chambers within the outer exine layer. The chambers within are crypt-like, the tectum being supported by rods sited on a floor layer. In pollen of *Hibiscus* in the family Malvaceae, the floor layer is enormously thickened.

The exine is made of *sporopollenin* which is remarkable for its resistance to both physical and enzymic degradation, and is believed to be produced by the oxidative polymerization of carotenoid pigments and carotenoid esters. The exine covers the pollen grain, except at the germinal apertures where it is usually absent or much reduced. It is present in most terrestrial plants but may be reduced to a thin membranous structure or even absent in some aquatic monocotyledons.

The inner wall layer, the intine, is smooth and does not contribute to pollen patterning but is exposed to the surface at the germinal apertures where it may be thickened and more complex in structure. Like the primary cell wall of somatic cells, it consists of cellulose microfibrils and a matrix of hemicelluloses, pectic polymers and proteins. It is present in all pollens and is often enormously thickened. The intine is developmentally quite distinct from the exine: it is laid down much later – at the early vacuolate period (see § 2.3).

1.4 Pollen analysis and the history of vegetation

The exine layer is endowed with all the patterning attributes remarkable to pollen which may be specific to family, tribe, genus or species. It is possible to relate pollen fossils preserved in geological deposits to living plant pollens (Fig. 1-6), providing valuable clues to the history of plant life in past geological eras. Fossil pollen deposits are dated by use of standard palaeontological procedures. In the case of pollen preserved in the relatively recent Quaternary peat deposits, radio-carbon dating is used. Measurements can be related to age within the past 30 000 years with some accuracy. Sediments which are geologically much older, are deposits of materials that originated elsewhere but have been deposited at the sampling site. Pollen grains within the sample can be extracted, identified and counted. The data, when assembled into pollen diagrams, gives an impression of the relative frequency of the genera or groups in the flora of the period, subject to obvious limitations that only the airborne pollen of wet environments, such as swamps, bogs and lake beds, tends to be preserved in large numbers in the fossil record. Pollen is preserved in such environments because they provide anaerobic conditions restricting oxidation of the sporopollenin. Not only do the results provide information about the vegetation of a particular area, but also about climatic changes that have occurred. An excellent review of the history of the British vegetation has been given by GODWIN (1975) and for central Europe by STRAKA (1975).

The principles used to identify fossil pollens are particularly interesting. Palynologists depend upon matching unknown fossils with present day reference pollens, making the important assumption that there have been no evolutionary changes in exine structure. The older Tertiary fossils present problems, since some are of extinct types, while others can still be related to a specific lineage (Fig. 1–6). For purposes of classification, fossil grains are given their own nomenclature, with their own generic and specific epithets. The antarctic beech, *Nothofagus*, becomes in fossil form *Nothofagidites*. The mimosa, *Acacia* becomes *Polyadopollenites*, named after its polyad pollen structure. In Australia, fossil eucalypt-like pollen is called *Myrtaceidites eucalyptoides* and has proved to be almost indistinguishable from other modern genera of Myrtaceae such as the apple myrtle, *Angophora*, the turpentine tree, *Syncarpia*, and the New Zealand pohutukawa or rata vine, *Metrosideros*.

The time ranges for various pollen grains in the fossil record are shown for the island continent of Australia in Fig. 1–7. Angiosperms first appear in the Lower Cretaceous, but it is not until the Upper Cretaceous that the first angiosperm pollen grains can be confidently identified. Many types of pollen make their debut in the Paleocene and Eocene, and fewer in the Oligocene and Miocene. The daisy family, or Compositae, makes its first appearance in the Miocene in Australia and elsewhere in the world. This huge family has worldwide distribution in nearly every ecological niche and has used almost every life form, yet it was apparently one of the last to evolve (Fig. 1–7). In Australia, one of the earliest pollen types to appear is that of the fossil *Nothofagidites*, which is recorded at low levels in the Paleocene and mid-Eocene, and in abundance in late Eocene and Miocene records. The late Miocene and Pliocene are notable for a fall in *Nothofagidites*, and are dominated by Myrtaceae. Later, in the Pleistocene, there is a rise in herbaceous plants such as grasses and Compositae.

These changes in vegetation can be related to what is now known of the history of continental geography – the movements of the continents over the earth's surface. Australia was connected to Antarctica until the Mid-Eocene time, which may account for the abundance of antarctic beech, *Nothofagus*, in the pollen assemblages until this period. Subsequently, Australia came into collision with the South-East Asian plate in mid-Tertiary, and changes are reflected in the fossil record, which provide

Fig. 1–7 Appearance of fossil angiosperm pollen during the Pleistocene period in Australia. (Adapted from MARTIN, H. (1978). *Alcheringa*, in press.)

important evidence for transcontinental migration of vegetation. H. Martin in Sydney recently considered that two forms of migration have occurred: (i) dispersal under favourable conditions; and (ii) climatically-controlled migration. In Australia, there is palynological evidence for a withdrawal from the central region, now desert, to the more climatically-favourable east coast; and also for a northward dispersal or withdrawal towards Queensland of present-day subtropical plants; and finally a withdrawal southwards to Tasmania seen, for example, in the present occurrence of the Antarctic beech, *Nothofagus gunii*. Fossil pollen data provides important evidence for such migration and the climatic control of plant evolution.

2 Formation of Pollen

Pollen is an integral part of the life cycle of flowering plants which is founded upon an alternation of two generations, as in lower plants. The dominant generation is the diploid vegetative plant body – the sporophyte. It bears haploid spores, single cells that produce either male or female gametes within structures called gametophytes. The pollen grain is the male gametophyte. At maturity, it may contain two, or in some cases three, cells. In about two thirds of flowering plant families, the pollen grains are shed from the anther in the two-celled condition (Fig. 2–1). One is destined to divide after germination to produce the two sperm cells, and the other to regulate pollen function. In other families, the sperm cell division occurs in the maturing pollen grain so that the pollen grains are three-celled (Fig. 2–1).

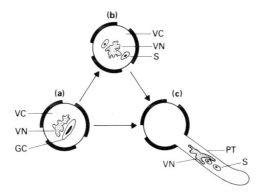

Fig. 2–1 Diagram showing mature binucleate (a) and trinucleate (b) pollen grains. The binucleate grains become trinucleate during germination (c). VC, vegetative cell; GC, generative cell; VN, vegetative nucleus, S, sperm cells. (Adapted from DEXHEIMER, J. (1970). *Rev. Cytol. et Biol. vég.*, 33, 169–234.)

Pollen is formed within the anther, which shows considerable variation in form but typically is an elongate structure containing four pollen sacs or loculi. The anther wall contains four layers (from outside): the epidermis, middle layer, tapetum and sporogenous cells (Fig. 2–2). The epidermis and middle layer are important for protection of the developing anther. The middle layer usually differentiates into the endothecium, with massive bar-like wall thickenings at maturity which are concerned with anther dehiscence. The tapetum is a nurse tissue surrounding the sporogenous cells, and there are two characteristic forms: parietal or secretory types in which the cells remain in position around the anther cavity; and amoeboid or plasmodial types where the cells become invasive, their cell walls breaking down so that the cytoplasm and nuclei make intimate contact with the developing microspores.

2.1 Meiosis and its consequences

Meiosis comprises two periods of cell division, which have important genetic consequences.

(i) Reduction of chromosome number by half, allowing for subsequent fusion of gametes to restore the original number.

(ii) Genetic recombination resulting from crossing over of chromosome segments ensuring that every microspore is different from its sibling, and that new variation may be expressed in the progeny.

The cytology and genetics of meiosis are well established (KEMP, 1970) and it is the cell biology of the events that is new and exciting. Much of the pioneering work in biochemical and electron microscopic aspects has been carried out by J. Heslop-Harrison, now at the Welsh Plant Breeding Station, Aberystwyth, and H. G. Dickinson at the University of Reading, England. They consider that meiosis 'is correlated in time with one of the most dramatic developmental changes known in plants, the transition from the sporophytic to the gametophytic phase of the life cycle. The sporophyte is the more elaborate of the two generations in morphology and biochemical capacities, and no doubt we see in it the expression of the greater part of the potentialities of the genome. The male and female gametophytes possess an independent metabolism, but they are highly reduced in morphology and range of functions, suggesting that a large part of the genome is repressed during this phase of the cycle. This repression must be imposed by some event – or sequence of events – occurring in the meiocytes or in the early life of the spores they produce. Furthermore, diplophase information which might interfere with the development of gametophytic functions must presumably be expunged during the same period before the cytoplasm becomes host for the gametophytic nuclei after the quadripartition of the meiocyte.' This is achieved by changes in the cellular organelles, reflecting a reprogramming or reorganization within the gametophytic cells; and by changes in structure and permeability of the cell walls during the sporophyte-gametophyte transition that initially enhance then inhibit communication between them.

The changes begin during early prophase of meiosis I, where in the anthers of lily there is a fall of up to 50% in total cellular RNA by pachytene, the period when chromosome pairing and crossing over becomes cytologically detectable. The adenine/guanine base ratio of the RNA alters, indicating a change from ribosomal RNA to a residual RNA associated with the chromosomes. Electron microscope studies of the pollen mother cells have revealed the elimination of a major part of the cytoplasmic ribosomes during prophase, so that by the end of first division virtually no ribosomes are detected in the cytoplasm. Ribosome levels are subsequently restored from cytoplasmic nucleoloids, nucleolus-like bodies, during second division. During the period of ribosome degradation, considerable portions of the cytoplasm become encapsulated within double- or multi-membrane systems, commencing in late-leptotene. The membranes have their origin in the nuclear envelope, isolating portions up to 16% of the pre-meiotic cytoplasm just before ribosome elimination. Within these inclusions, the ribosomes remain apparently unaffected by the degradative processes outside. Changes occur in other cytoplasmic organelles, the plastids and mitochondria undergoing a reorganization of internal structure.

While the nuclear and cytoplasmic events of prophase are in progress,

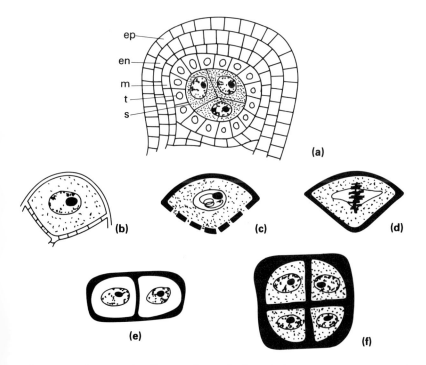

Fig. 2-2 Diagram of pollen mother cells (PMC) during meiosis. (a) TS of portion of four-layered anther wall showing pre-meiotic PMC surrounded by tapetal cells. (b) Detail of single PMC from (a) showing plasmodesmata connecting neighbouring cells. (c) PMC at zygotene-pachytene. Primary cell wall of (b) is replaced by a callose wall penetrated by large channels. (d) PMC at metaphase I. Callose wall continuous. (e) and (f) Dyad and tetrad. Young microspores separated and surrounded by callose wall. ep, epidermis; en, endothecium; m, middle layer; t, tapetum; s, sporogenous cells.

equally dramatic changes occur in the cell walls and intercellular connections (Fig. 2–2). The pollen mother cells, when first differentiated from the sporogenous cells, are separated by primary cell walls, composed of cellulose microfibrils within a matrix of hemicellulose and pectins. The plasma membranes of neighbouring cells are connected laterally via plasmodesmata, narrow tubules in the wall. During early prophase, the primary walls are degraded and replaced with walls made of the 1,3 β-glucan, callose, penetrated by massive cytoplasmic channels, each 1 μm in diameter, making up 20% of the wall interface, and reaching their maximum development at zygotene-pachytene. At this period, the entire mass of cells is a syncytium that may be withdrawn with a needle from the anther cavity as a coherent structure. This freedom of cytoplasmic communication is probably related to the requirement for meiotic synchrony, all the cells progressing through the various sequential periods together.

Ultimately, the callose walls of the pollen mother cells close together, sealing the cytoplasmic channels by first metaphase. Subsequently, a special wall of callose invests the dyad and tetrad configurations to a varying degree in different families. The function of the callose wall is probably to provide a degree of isolation, to preserve the genetic identity and integrity of the developing microspore, since callose has been found to retard entry to the protoplast of macromolecules larger than simple sugars. Alternatively, it has been proposed that callose being hygroscopic can absorb large amounts of water, and would prevent undesirable desiccation of the sporogenous cells within the anther under conditions of water stress. Whatever its function, by the end of the tetrad period, the young microspores are freed from the callose wall by the action of 1,3 β-glucan hydrolase enzymes which digest the wall away. The presence of callose investing the meiotic cells or tetrads of microspores can be readily demonstrated cytochemically (see Chapter 7).

2.2 The puzzle of wall pattern

The earliest stages of pollen wall development, including the formation of the exine, occur while the microspores are embedded within the callose wall at tetrad period. When formed, the microspores within the tetrad have protoplasts surrounded by a plasma membrane and the callose wall, and no patterning is evident. However, the future sites of the germinal apertures, which are related to the position of the microspore within the tetrad, are apparently already determined. Sheets of endoplasmic reticulum membrane (see § 2.4) have been observed lying adjacent and parallel to the plasma membrane only at these sites. The *primexine*, a layer which provides a matrix within which the pattern elements of the exine are laid down, appears next. It is composed of a microfibrillar cellulosic material, and is situated between the plasma membrane of each microspore and the callose wall. It is absent at the sites of the spore

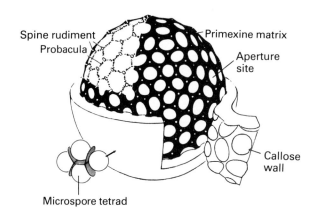

Fig. 2–3 Cutaway diagram of *Ipomoea* microspore at tetrad period. Each microspore is surrounded by a callose wall, overlying the primexine matrix (black). The spine rudiments and probacula or exine (white) lie within the primexine matrix. (Adapted from WATERKEYN, L. and BIENFAIT, A. (1970). *Grana*, 10, 13.)

apertures (Fig. 2–3). The rods of the outer exine are defined during primexine formation. In lily microspores, Heslop-Harrison has shown by transmission electron microscopy that the sites destined to become rods are marked by the appearance of columns of radially-oriented convoluted lamellae within the primexine layer. When synthesis of sporopollenin begins, the lamellae of the columns are coated, and their appearance changes to that typical of the rods. At the same time, lamellae form below tangentially along the surface of the plasma membrane, and these are coated and become the floor layer of the exine. In tectate grains, the spines that will later ornament the exine are also formed, but remain soft and flexible, while within the tetrad. The result is that by the late patterned period all the elements of exine patterning are present in the spore wall. At spore release, the callose wall breaks down, releasing the microspores.

The primexine matrix material becomes dispersed during the rapid exine growth that is typical of the young spore period. Freed of the constraints of the callose wall, changes in shape also occur, the grains becoming spherical. Accumulation of sporopollenin for exine growth continues during the active life of the tapetum, but involves only the elaboration of structural features present at spore release.

There is good reason to assume that the tapetum plays an important role in build-up of the exine and pollen nutrition. At the young spore period, the tapetal cells have all the characteristics of secretory cells. The tapetal cell wall adjacent to the pollen sac changes in appearance, and granules of sporopollenin accumulate at the tapetal cell surfaces. These structures, known as *orbicules*, were first noted by nineteenth-century cytologists in several families of flowering plants, and are often termed *Ubisch bodies*. They grow in size as the spore exines increase in thickness by accumulation of sporopollenin from the external surface. Many functions have been proposed for the orbicules: transport form of sporopollenin; source of sporopollenin for exine deposition; forming a sac to assist pollen dispersal; or a by-product of anther metabolism. Whatever their function, they retain all the characteristics of the sporopollenin deposited in the exine, including some elements of patterning, and range in size between 6 to 8 μm.

The generation of these complex patterns raises questions of especial interest concerning the genetic control mechanism, both in terms of its location and the manner in which it operates. The exine is produced around the haploid spore, but within an anther environment that is entirely diploid and parentally-specified. The puzzle of the wall pattern is whether the control is regulated by the microspore or by the parental sporophyte. The information required to specify exine pattern must be borne by the spore, since the wall is patterned while the spore is sealed within the callose wall. However, evidence from genetic studies shows that the whole process is under sporophytic control.

W. Bateson and R. C. Punnett in 1909 carried out breeding experiments with two pure lines of sweet peas, one with long-shaped pollen (usually with three apertures) and another with round grains (with two apertures). When crossed, the F_1 plants all had long pollen, suggesting that long is dominant to round. The F_2 plants showed a 3 : 1 segregation of long to short pollen, confirming that the long character is determined by a dominant gene L, and round by the recessive allele l. Since all F_1 plants were long, it is clear that control over shape of the grains is retained by the sporophyte, as gametophytic control would have resulted in a 1 : 1 ratio. What is the executive agent? Can it be the tapetal cells around the spores during wall synthesis, or are cytoplasmic determinants handed down through the cell lineages of meiosis from the pollen mother cell? If the pollen mother cells are disrupted either mechanically by centrifugation or chemically with the drug colchicine, errors in pollen wall formation result. There are apparently two vulnerable periods: early prophase – in which treatment affects aperture location and grain separation, and the second meiotic division, early tetrad period – in which exine pattern defects may be induced. These experiments suggest that at least partial control is exerted by the sporophytic pollen mother cell.

2.3 Pollen development

Following spore release, the various periods of pollen development have been characterized by Heslop-Harrison and co-workers using morphological and chemical criteria. Spore release heralds the beginning

§ 2.3   17

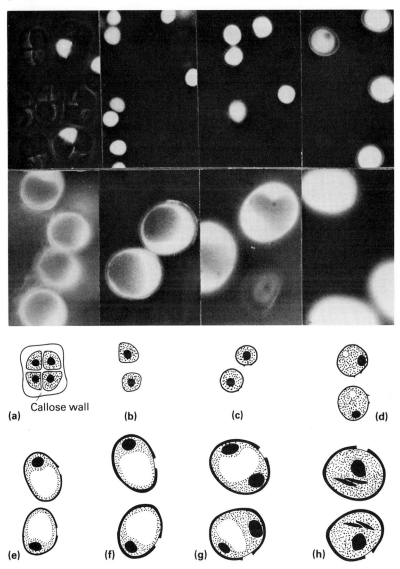

Fig. 2–4 Morphological features of pollen development shown by fluorochromatic reaction (FCR) in photographs are shown diagrammatically. (a) Microspore tetrad – fluorescence only in damaged grains. (b) Spore release – non vacuolate. (c) Young spore period – non vacuolate. (d) Early vacuolate period – vacuole increasing in size. (e) Mid vacuolate period – vacuole nearly fills grain. (f) Late vacuolate period – vacuole reduced in size. (g) Early maturation period – binucleate grain. (h) Mature trinucleate grain – cytoplasm filled with starch grains.

of the *young spore period*. The free spores rapidly increase in volume, are non-vacuolate and are bounded by a plasma membrane and thin exine (Fig. 2–4).

The major growth period of the pollen grain commences with the onset of the *vacuolate period*. It is signalled by the appearance of a small central vacuole within the cytoplasm of the grains, and by synthesis of the intine wall layer. The vacuole is an organelle bounded by a unit membrane called the tonoplast. The vacuolate period can be arbitrarily divided into three phases, the early- , mid- and late-vacuolate periods. In the early period, the vacuole increases in size relative to the cytoplasm till it is about half the grain diameter (Fig. 2–4), intine polysaccharide synthesis begins from within the grain, as does the laying down of the inner exine layer, the nexine. At mid-vacuolate period, the grain resembles a 'signet-ring' in appearance, with the vacuole now filling the grain except for a thin film of cytoplasm around the periphery of the plasma membrane and the prominent single nucleus at one end. The intine is now well-developed and is conspicuously thickened at the germinal aperture. By late vacuolate period, the grains have greatly increased in volume, but the vacuole has not kept pace and is restricted to about half the diameter of the grain (Fig. 2–4).

In most pollens, the vacuolate period ends with the first pollen mitosis. Until this time, the grain has technically been a microspore. This division produces a central *vegetative nucleus*, and a *generative cell* which will later divide again producing the sperm cells. The vegetative nucleus directs the functioning of the pollen grain and its pollen tube.

Pollen grain mitosis is an unequal division, the generative cell being markedly smaller than the vegetative cell. The vegetative nucleus enlarges during pollen maturation, often becoming markedly lobed in appearance. The generative cell is at first hemispherical and usually lies nearest the intine. It differs from a somatic cell in that a callose wall is deposited initially instead of cellulose. The callose wall is ephemeral, and within 24 hours the polysaccharide wall has been broken down and replaced by two ensheathing membrane layers. Cytoplasmic organelles are usually present in the generative cell including endoplasmic reticulum, mitochondria, plastids and ribosomes. These may be important in the transmission of male cytoplasmic determinants to the zygote at fertilization.

In binucleate pollens, the generative cell generally elongates, and enters what seems to be a suspended mitotic prophase. Division to form the two gametes does not occur until after pollen germination. In trinucleate pollens, such as grasses (Figs 2–4 and 2–5), division occurs during pollen maturation. On division, the two gamete nuclei are initially separated by a cell plate. This is broken down and replaced by bounding membranes so that the two sperms are effectively cells, even if much reduced. The sperm cells may contain a range of organelles, and assume an elongate or thread-like appearance.

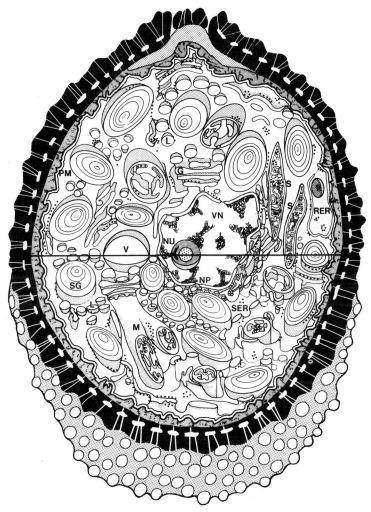

Fig. 2–5 Diagram of pollen grain of ryegrass cut to reveal the nuclei, cytoplasmic organelles and wall layers and their interrelationships. Not to scale, only a few examples of each organelle shown. Wall layers: exine (black) perforated by radially oriented micropores (white); intine (stippled) thickened at single germinal aperture (uppermost part of cell. Organelles: SER, smooth and RER, rough endoplasmic reticulum; M, mitochondria; G, Golgi apparatus; SG, starch grains; V, vacuole; L, lipid droplets; PM, plasma membrane; VN, vegetative nucleus; NU, containing nucleolus; NP, nuclear pores; and S, sperm cells surrounded by own plasma membrane; for simplicity the additional plasma membrane envelope from the vegetative cell has been omitted. (Based on electron micrographs and an adaptation of the plant cell diagram by GUNNING and STEER, 1976.)

The period following first pollen mitosis is called the *maturation period*. J. W. Woodard has shown with *Tradescantia* pollen that there is a fivefold increase in protein storage during the maturation period. The central vacuole has become dumb-bell shaped, and progressively reduced as cytoplasmic synthesis proceeds. It is generally useful to distinguish the late maturation period by the onset of carbohydrate storage in the grain which is very evident by the accumulation of large starch grains in the dense, non-vacuolate cytoplasm.

In the grass pollen grain, there is a 500-fold increase in volume between spore release and maturity. While the events described hold for most pollens where development has been investigated, interesting exceptions occur. These are mostly minor changes in sequence, but occasionally the whole developmental sequence is changed. This has been shown in the unusual filiform pollen of certain sea-grasses. The first microspore division occurs shortly after spore release, so that the vacuolate period commences after rather than before the first pollen mitosis. In most pollens, the duration of development is short, the interval between spore release and pollen mitosis being about 5 days in *Tradescantia*, corn and canary grass, 15 days in lily and more than 30 days in some orchids. In many woody plants, there may be a prolonged period of dormancy between meiosis and pollen maturation.

2.4 Cell biology of the pollen grain

The pollen grain behaves as a single cellular unit, even though the mature grain may consist of two or three cells. The ultrastructure of pollen is shown in Fig. 2–5, using the pollen grain of ryegrass as an example. The grain is monoporate, and three-celled. The largest cell, the vegetative cell, has a lobed central nucleus containing chromatin and a single nucleolus. The nuclear membrane has many pores communicating with the cytoplasm and connected with the membrane system of the endoplasmic reticulum. The nucleus lies within the cytoplasm which is densely packed with storage organelles: carbohydrate is stored in starch grains, lipids in spherosomes, and protein in membrane vesicles associated with the endoplasmic reticulum. The organelles involved in the biosynthesis of these storage products are apparent: Golgi apparatus for carbohydrate synthesis, smooth and rough endoplasmic reticulum for protein synthesis. The cytoplasm is also packed with mitochondria and plastids which provide the metabolic requirements of the cell (see GUNNING and STEER, 1976, for review). These storage granules are mobilized during germination and sustain the growing tube until it is in contact with the style tissue, from which it may derive further nutrition.

The other two cells are the sperm, which lie within the vegetative cell. Binucleate grains contain a single generative cell which is similar but larger. Earlier cytologists, using the light microscope, had not agreed whether the sperm were cells or naked nuclei. This was resolved by D. A.

Larson in 1965 who showed by electron microscope observations that the sperm of corn, *Zea mays*, have a limited cytoplasm and are surrounded by a double membrane, and so are true cells. Each sperm is an elongate structure about 8×14 μm, surrounded by a distinct plasma membrane, and, in addition, by a plasma membrane from the vegetative cell. The sperm nucleus is spherical with a porous nuclear membrane. Organelles present in the cytoplasm tend to be reduced in internal structure compared with their counterparts in the vegetative cell, and include Golgi apparatus, mitochondria, endoplasmic reticulum vesicles and polysomes. Plastids have also been identified in some pollens.

The plasma membrane of the vegetative cell is surrounded by a two-layered wall. The inner intine layer contains many radially-aligned tubules of proteinaceous material, and is characteristically thickened at the single aperture or pore by an intermediate layer or *zwischenkörper*. The outer exine layer is tectate, and both the nexine, or foot layer, and the outer sexine are traversed by fine channels, the *micropores*.

The outer surface of mature pollen is usually coated with a sticky, pigmented, lipid-rich material named *pollenkitt* (pollen cement), by F. Knoll in 1930. It is a secretion from the tapetal cells and its adhesive properties stem from its oily, proteinaceous nature, and its colour from carotenoid pigments. It is present in particular abundance in the pollen of animal-pollinated plants, presumably to enhance adhesion to the animal's body. In addition, viscin threads may form strands between grains, as in *Rhododendron* in the family Ericaceae, and *Oenothera* in the family Onagraceae. In these anthers, the grains are linked by the thread and released as a string of pollen. In orchids and asclepiads, the entire contents of the pollen sac may be glued together to form a pollinium (see p. 7). The pollenkitt may not only function as an adhesive, but may also protect the sperm from ultraviolet (UV) irradiation by virtue of its UV-absorbing pigments. It may also contain attractants for pollinators such as scents. Gibberellin-like growth hormones known as brassins have been found in the pollenkitt of rape and crucifers.

When ready for dispersal, pollen grains are enclosed within their protective walls in a state of temporary dormancy. In the 1920s, H. E. Knowlton and C. G. Vinson demonstrated their extraordinary low respiratory and metabolic rate. The water content of the pollen is reduced to 10–15%, similar to levels in seeds. However, pollen dormancy is dependent on the state of hydration rather than on the presence of any inhibitors which may regulate seed dormancy.

3 Dispersal of Pollen

The anthers present their pollen for dehiscence in two ways. Most commonly, the anther splits longitudinally. Alternatively, the pollen may be released by a trap mechanism, through a terminal pore, lid or valve; in the family Ericaceae, the trap is operated by the pollinating insect which squeezes the anthers of *Rhododendron*, or strikes the anthers of *Vaccinium* or *Thunbergia*. There is no dehiscence mechanism in some genera, such as the mangosteen *Garcinia* in the family Guttiferae, where the pollen has to be forcibly removed by the insect pollinator.

Secondary pollen presentation occurs when the pollen is emptied from the anthers which dehisce within the closed flower, and is deposited at some other site within the flower. In *Lechenaultia*, in the family Goodeniaceae, the pollen is held within a purse-like structure which, after the pollen is collected by vectors, becomes part of the stigma. In the waratah, *Telopea*, and other Proteaceae, the pollen is captured at the tip of the stigma column, and the underlying stigma becomes receptive only after the pollen has been removed by bird pollinators (Fig. 3–3). In the harebell, *Campanula* (Campanulaceae), and sunflower (Compositae), sterile hairs on the elongating stigma push the pollen clumps out of the anthers and deposit them outside the floret, thus facilitating collection by pollinating insects. This process is taken to the extreme in *Centaurea*, a member of the thistle tribe, where pollen exposure will only occur if the filaments are irritated by the visit of a pollinating insect, when they contract and expose the pollen. It has been suggested that this is a mechanism to prevent the theft of pollen by insects, or its being blown or washed away.

3.1 Adaptations for wind dispersal

Only about 30 out of more than 300 families of flowering plants show adaptations for pollen dispersal in air currents, termed *anemophily*. Families in which most of the genera are wind-pollinated include: Gramineae (grasses), Cyperaceae (sedges), Juncaceae (rushes), Salicaceae (poplars), Chenopodiaceae (dock), and Urticaceae (nettles). Among several families that are predominantly pollinated by animal vectors, some genera are wind-pollinated, for example, the ragweed, *Ambrosia*, in the Compositae and the ash, *Fraxinus*, in the Oleaceae. It is a feature often associated with separation of the sexes in the same or different plants: monoecious (Greek: one house) ragweed and birch; dioecious (Greek: two houses) oak and poplar.

Wind-pollinated flowers show adaptations towards reduction of the floral parts, increased pollen production and size of stigma, together with a more effective presentation of both organs. In the case of the anthers, the long filaments place them well outside the open perianth, and they are loosely attached to the filaments so that they are versatile and may vibrate and move in air currents to aid pollen release. The anthers of wind-pollinated species open only during favourable weather when it is warm and dry. Consequently, they tend to open during daylight hours only. Some grasses show a bimodal pattern of flower opening, both in early morning or late afternoon, while others have only one daily flowering period.

Flower opening in rye can be experimentally induced in plants ready to flower by shading the plants for at least ten seconds. Buus Johansen in Denmark showed in 1966 that a burst of flowering followed after a time lag of only two minutes. He concluded that a rapid decrease in incoming radiation, such as could be produced in nature by the passing of clouds, could be the general trigger for the mass flowering of grasses. Flower opening is induced by the lodicules, small white petal-like structures within the flower, which swell rapidly at the base, forcing the green lemma outwards. At the same time, the filaments of the stamens suddenly extend by cell elongation from about 2 mm in length to a final 7–10 mm, at a rate, in rye, of 1.6 mm per minute. Dehiscence is considered to occur by withdrawal of water from the anther endothecium, causing it to split open and release the pollen. The flowers remain open for 5–15 minutes, and then the glumes slowly close. The entire process was delightfully described by Zuderell in 1909 (cited from ARBER, A. (1934), *The Gramineae*, Cambridge University Press, p. 159). Rye spikes kept in a darkened room were suddenly exposed to sunlight: 'in the next half-minute, a peculiar stir occurs, a delicate crackling; the glumes begin to relax their connection. . . . Here and there, up between the glumes, the tips of the pretty violet anthers peep shyly out, and forthwith a general commotion begins; in each flower a contest appears to arise between the sister anthers as to which can first escape from their confined quarters! The frail delicate filament, which is elongating more and more, can no longer support the weight of the anther; it tips over, and the others follow, scattering around themselves little dust clouds of pollen grains. A strange twisting and quivering seems to agitate them; they become flattened out, right to the apex. More and more anthers emerge, and the pollen clouds become more numerous and larger, until finally the powdering is universal; millions of pollen grains cover the table around the jar.'

Massive pollen release can be observed on a still morning in a field of grass. When the first breeze stirs the flowers, the pollen rises in small clouds forming a haze over the field. If not mown or grazed, a one hectare field of ryegrass will release an estimated 210 kg of pollen in one flowering season. Pollen output of different grasses varies widely, the common agricultural grasses such as ryegrass, cocksfoot, yorkshire fog and canary releasing between two and five million pollen grains from each flowering spike; while others, such as brome and wild oat, release less than 1000 grains.

3.2 Pollen transport by insects and other invertebrates

Animal pollen dispersal involves very specific co-adaptation between the flower and its pollinator. The pollen itself is generally coated with an adhesive to ensure it will stick to the body of the insect. It is dispersed in clumps, rather than free grains, to provide a convenient size of load for the pollinator. This is carried to extremes in certain families where special mechanisms have evolved to ensure efficient dispersal; for example, in *Rhododendron* (Ericaceae) the glue coating is reinforced with long viscin threads which ensure that the pollen is released in long strings; in certain orchids and asclepiads, the pollen adheres in masses or in special structures (pollinia) which form single pollen loads.

In order to establish direct and frequent relationships between flower and pollinator, special attractants are provided, including the supply of nectar, pollen or a special scent. Flowers may also be visited by several different kinds of pollinators. In the absence of such direct attractants, pollination can only occur as the result of accidental visits. Earwigs and grasshoppers are claimed to have pollinated sunflowers during the night as a result of a chance visit to the flowers.

A list of insect pollinators is given in Table 1. Primitive insects mess around in the flower, becoming more or less covered in pollen over their legs and body, some part of which may touch a stigma. This is the case with the thrip, *Taeniothrips ericae*, which may pollinate heather, *Calluna vulgaris*. The bodies of the thrips become covered in nectar, and the pollen tetrads adhere to them. Males are uncommon, so females move from flower to flower, transferring pollen as they search for a mate.

Beetles are important in pollination of tropical flowers, and show special adaptations for pollen chewing. *Nemognathus*, for example, has maxillae which usually are longer than its body. The beetles are attracted to flowers by odour. Members of the tropical family Annonaceae give off a fruity odour, while *Calycanthus* smells of fermenting fruit. The beetles feed on the sticky pollen and in the process may carry out pollination. *Calycanthus occidentalis* (allspice) has a trap device at the base of the flower, covered by spirally-arranged rows of red tepals which bend inwards like a lobster pot covering the anthers and pistil. Its fruity odour attracts the beetle, *Colopterus*, which enters the trap where it eats special fruit bodies formed from sterile anthers. Once inside, exit is barred by the reflexed tepals, trapping up to ten beetles inside each flower. Initially, the stigmas are receptive, but wither before anther dehiscence, which does not take place until two days after flower opening. The beetles are then covered in pollen, and the flower opens, setting them free to visit newly-opened flowers and carry out cross-pollination.

Similar trap mechanisms are used to attract carrion or dung flies,

§ 3.2 POLLEN TRANSPORT BY INVERTEBRATES

especially for the pollination of aroids, which may produce an odour of decaying faeces. The more primitive types of flies merely seek nectar in odoriferous flowers such as ivy, *Hedera helix*, and various Umbelliferae, while in the more highly adapted flies the proboscis is greatly elongated for nectar feeding, the bumble-bee-like Bombyliidae have prosbosces approaching 10 mm long, the tabanids and nemestrinids of southern Africa have prosbosces exceeding 50 mm in length. Since flies do not require pollen to nurse their larvae, their pollinating activities are irregular.

The *Hymenoptera* include some of the economically most important pollinating insects. Of these, only the wasps and bees feed their larvae on pollen (Table 1). Among the imagines (adults), only the sawflies and bees eat pollen as a source of protein. Ants are well-known nectar eaters, and

Table 1 List of pollinating insects and their adaptations.

	Adaptation				
	Larvae feed on		Imagines feed on		
Insect	nectar	pollen	nectar	pollen	flower odour
THYSANOPTERA					
Thrips	–	–	+	–	–
COLEOPTERA					
Various beetles	–	–	+	+	–
Dung beetles	–	–	+	+	+
DIPTERA					
Various flies	–	–	+	+	–
Carrion or dung flies	–	–	+	+	+
HYMENOPTERA					
Symphyta: sawflies	–	–	+	+	–
Terebranthes, Ichneumonidae:					
ichnueumon wasps (floral mimicry)	–	–	+	–	–
Aculeata, Formicidae: ants	+	–	+	–	–
Vespidae: wasps	+	(+)	+	–	–
Apioideae: bees	+	+	+	+	–
LEPIDOPTERA					
Rhopalocera: butterflies (diurnal, alight on flowers)	–	–	+	–	–
Heterocera: moths (nocturnal)					
Noctuids (alight on blossoms)	–	–	+	–	+
Sphingids (hover)	–	–	+	–	+

less well-known as pollinators; the familiar house-plant of Victorian times, *Aspidistra*, is pollinated by ants while feeding on its copious nectar in the purple flowers hidden under the plant at soil level. In tropical deserts, such as the Sahara, ants are important in pollinating several Euphorbiaceae. These have extra-floral nectaries easily accessible to ants, which are one of the few insects able to tolerate the extremes of temperature.

The most curious pollination adaptation of all must surely be that of certain male hymenopterans, such as the ichneumon wasp, which pollinates certain orchids. The flowers attract the male wasps by mimicking the form of the female insect. Young male insects, who have usually not encountered a female, are attracted to the flowers, and in the course of attempting to copulate, the insect touches the rostellum of the flower with its body and carries off the pollinia (Fig. 3–1). It will then proceed from flower to flower repeating the performance, and

Fig. 3–1 Pollination of the Australian small tongue orchid, *Cryptostylis leptochila*, by the ichneumon wasp, *Lissopimpla semipunctata*. The wasp (a) entered the flower backwards with its abdomen towards the stigma (b). The flower's labellum is gripped by its claspers, the insect quivered for a moment, then became motionless before freeing itself from the flower with a pair of pollinia attached to its abdomen. The wasp had visited three flowers in rapid succession. (From COLEMAN, E. (1930). *The Victorian Naturalist*, **46**, 237.)

effecting cross-pollination. The orchid pollinia are always deposited in a special place on the visitor's body; the ichneumon wasp in Fig. 3–1, visiting *Cryptostylis leptochila*, finds them attached to its abdomen; *Platanthera chlorantha* places them on the insect's eyes; *P. bifolia* places them on its proboscis; and *Sabia pratensis* on its thorax. The wasp may then carry the pollinia for long distances to another flower.

Male bees may be attracted to other orchid flowers because they perceive them as if they were enemy insects, intruders into their territory. The UV-reflecting central region of the flower, the nectar-guide, mimics the appearance of enemy insects, and during the aggressive response, pollination may be effected. The pollen of orchids, because it is transported *en masse*, is not available for food, so that the flowers have to decoy polliniverous insects such as bees by providing attractive nectar.

Bees are the most highly evolved of all the insects for pollination. They exhibit a wide range of behaviour from the relatively simple pattern of the solitary bee to the complex patterns of the hive- and social-bees. Many solitary bees, for example in the *Prosopididae*, have short mouth parts and eat the pollen which is later regurgitated for the brood. The leaf-cutter bees of the *Megachilidae* collect pollen ventrally, while all higher bees are foot-collectors. The collecting sites are the hairy feet in the lowest types, to the corbiculae or pollen sacs on the hind legs in the most advanced bees. The bigger bees are strong insects, able to move floral parts about while seeking out pollen or nectar. Flowers such as *Delphinium* and *Aconitum* which have open flowers with long spurs are exclusively bee-pollinated while the closed flowers of many Labiatae, Papilionaceae and Scrophulariaceae are pollinated by bumble-bees. These big bees can carry considerable quantities of pollen, and even if they attempt to comb out all the pollen sticking to their fur, enough adheres for pollination. W. C. Adler in 1966 showed that eight visits by bees are needed for good seed-setting in water-melon flowers.

Honey-bees visit flowers that may provide pollen, or nectar, or both. Dorothy Hodges, an English apiarist, described in 1952 how she 'watched the procession of pollen-laden bees returning from the field, their pollen loads either sombre with brown or black, or glowing bright colours, orange, yellow, red or green.' The bees mix the pollen into a paste with nectar from their honey sacs to form the pollen load. The procedure is a fascinating one. The bee cleans the pollen from its body with its tongue and legs, transferring it to the brushes of the middle legs. The process is illustrated in Fig. 3–2. The tips (metatarsi) of the hind legs are held together, while those of the middle legs are placed in turn between them and drawn forward. As Hodges described it: 'this leaves the pollen, now in a sticky mass, loaded into the rows of stiff combs on the inner side of the hind metatarsi. When sufficient pollen is collected here, the final action takes place. At the distal end of the inner side of the tibia of the hind legs is a strong rake . . . which is held against the top of the opposite metatarsal combs and pushed downwards, thus raking out all the moist pollen in a compact mass into the pollen press between the rake and the auricle of the opposite leg. The joint between the tibia and the tarsus is then closed and the paste-like pollen is squeezed in the pollen press and pushed outwards

Fig. 3-2 Bees working the catkins of willow. Bee on the left is using her mandibles to scrape pollen from the anthers; bee on the lower right is transferring honey from her tongue and loading her combs; bee flying, on the upper right, is using her middle legs to shape the load. (From HODGES, 1974.)

and upwards . . . it comes to rest on the smooth concave floor of the corbicula or pollen basket.'

On returning to the hive, the pollen is stored in empty or partly filled wax cells. The bee lowers her hind legs into the cell and squeezes out the pollen with her middle legs. Another bee will press the pollen firmly down and, when full, seal the load with honey for storage. Each cell holds about 15–20 loads, each of 7–15 mg of pollen. Pollen stored in this way is called bee-bread. During storage, it may undergo biochemical changes, including increased acidity, inversion of pollen sucrose, and develop a high histamine content. Many of these changes may be due to microbial contamination, and a specific bacterial flora is associated with bee-bread.

The kinds of pollen stored vary depending on the flower season. In Europe, E. P. Jeffree and M. D. Allen in 1957 showed that tree pollens are first to be collected in early spring (February–March), especially hazel, poplar, *Viscum* and willow. In late spring (April–May), pollens of fruit trees, dandelions and crucifers, together with oak and ash trees, predominate. In early summer, clover, mustard, comfrey, poppy and other weeds are frequent. Most of these plants present their pollen in late morning and afternoon, between 10.00 and 16.00 hours daily. Some, however, open only in the mornings between 06.00 and 10.00 hours including roses, poppy, ragwort, *Verbascum* and *Verbena*.

The more primitive lepidopterans have chewing mouthparts and eat pollen, while the higher forms are exclusively nectar feeders using their long thin proboscis to seek it out. Butterflies visit flowers during the day and alight to seek out nectar and pollen. In contrast, moths are mostly nocturnal, and hover in front of the flower (Table 1). All food collected is consumed by the imagine itself as none feed their larvae. Moths have a strong sense of smell, perhaps an essential adaptation for a nocturnal pollinator, and can also pollinate extremely long flowers. Hawkmoths may compete with humming-birds in pollinating certain flowers. Lepidopteran-pollinated flowers are usually more delicate, often with versatile anthers, in contrast to the tough flowers with fixed anthers that are typical of those pollinated by birds.

3.3 Pollination by birds, bats and mammals

Vertebrates have much more complex and greater food requirements than invertebrates, but both birds, bats and some small mammals eat pollen and nectar sometimes as a major part of their diet, and may act as pollinators. The sugar content of both pollen and nectar is probably the major attractant, providing the source of energy for the very high metabolic rate of humming-birds, which need $1.75 \text{ J g}^{-1} \text{ h}^{-1}$ and eat twice their own weight each day. Since vertebrates have greater longevity, they demand food all year round, so that vertebrate pollinators are generally found only in more tropical regions, where flowers occur in all seasons.

Birds often obtain nectar by puncturing flowers. Indeed, there is an Indonesian genus of Loranthaceae where the flowers remain closed until they are punctured by a bird and then they open explosively. Birds that carry out pollination include the New World humming-birds, the African and Asian sun birds, the Hawaiian honey-creepers, American sugar birds, Australian honey-eaters and lorikeets. Lorikeets are also known as brush-tongued honey parrots, the name indicating the fringed tip of the tongue which is a characteristic of birds adapted for nectar feeding. While humming-birds perform their activities hovering in front of the flower, others such as the honey-eaters perch on the flowers or the branch beneath them. Flowers adapted for bird pollination have to be strong enough to both provide a perch for the bird, and withstand damage from penetration by its beak. Bird pollination is a feature of many Australian plants, whose flowers often exhibit a pincushion style of inflorescence to meet these requirements, and are frequently red, a colour that is particularly attractive to birds. The waratah, *Telopea*, and many species of silky oak, *Grevillea* (Proteaceae), have the pollen squeezed from the anthers onto the tip of the stylar column when the flower opens. Honeyeaters arriving on the inflorescence seeking out nectar may have pollen dusted on the top of their heads (Fig. 3–3). On moving to an older flower and seeking nectar again, the pollen then touches the now receptive stigma at the same site.

Fig. 3–3 An Australian spinebill honey-eater about to pollinate a flower of *Grevillea*. The bird has received a cap of pollen on its head while feeding from another flower and when its long beak enters the flower completely, the head contacts the stigma at the end of the column effecting pollination. (From MORCOMBE, 1968.)

M. W. Burck made the first observations of bat pollination during a visit to the Bogor Botanical Garden in Indonesia in 1892. He observed fruit-eating bats visiting flowers of *Freycinetia insignis*, and are now known to be entirely bat pollinated. Bat-pollinated species include the banana, *Musa*; the baobab, *Andansonia*; the durian, *Durio*; and the bread-fruit, *Artocarpus*. Bat flowers usually give off a characteristic odour that is said to resemble that of the animals themselves, reminiscent of butyric acid; banana flowers have a musty odour. Bat flowers are often borne on long hanging stalks, which in the woody climber, *Mucuna*, may be up to 10 m in length. Cauliflory, where the flowers are borne directly from tree branches, is also a characteristic adaptation for the mode of pollination since bats like to hang upside down while feeding. Bat flowers produce large quantities of pollen, the baobab, for example, has 1500–2000 anthers within a single flower. Nectar- and pollen-eating bats belong to the sub-family Glossophaginae, which all possess a longer more tapering snout and tongue than their insectivorous relations. *Musonycteris harrisonii* has a tongue that is as long as its body. S. Vogel in 1968 showed that the hairs of *Glossophaga* are similar to the scales on the abdomen of the bumble-bee, and are apparently an adaptation for pollen transport.

Many small vegetarian mammals eat flowers and may pollinate them. O. Degener in 1945 noted that the Hawaiian night rat, *Rattus hawaiiensis*, climbed trees of *Freycinetia arborea* to eat the succulent bracts around the flowers and transferred pollen while feeding. The baobab, *Adansonia digitata*, may be pollinated by the bush baby, *Galago*. Marsupials, such as the honey mouse, *Tarsipes spencerae*, and the Australian endemic rat, *Rattus fuscipes*, regularly pollinate the pincushion flowers of several Proteaceae, including coast or slender *Banksia attenuata*, while feeding on the nectar.

3.4 Underwater pollination

Water currents provide as ideal a medium as the air for pollen dispersal of aquatic plants. However, since terrestrial pollen will usually burst if placed in water, special devices are used by aquatic plants to waterproof their pollen or adapt it for the freshwater, brackish or marine environment. Many aquatic plants inhabiting all but marine habitats overcome the problem simply by flowering above the water surface. In an analogous way, some terrestrial pollens are claimed to be pollinated by raindrops, enabling pollen to float from the anthers to the stigma, and by rain-splash, to be transferred from flower to flower. Other aquatic plants are submerged throughout their life cycle, but manage to carry out pollination on the water surface.

One of the most effective mechanisms is bubble pollination, where the male flowers, anthers or even pollen are released from the submerged plant within a bubble of air and float to the surface for pollination. In eel grass, *Vallisneria* of the family Hydrocharitaceae, J. Scott in 1869 observed the spectacular events: 'under a noonday sun, the innumerable florets freed from their spathes and ascending like tiny air globules till they reach the surface of the water, where the calyx quietly bursts – the two larger and opposite sepals reflex, forming tiny rudders, with the third and smaller recurved as a miniature sail . . . facilitating in an admirable manner the floret's mission to those of the emerging females.' The female flower, at first submerged, is rapidly borne to the surface by the elongating peduncle, which appears coiled like a spring.

Another member of the same family, the sea-grass *Halophila*, has adopted a different approach. Its spherical pollen grains are released into the sea in long sticky threads, aiding and increasing the probability of contact with the stigmas. These observations were first made by the French explorer and botanist Gaudichaud at Shark Bay, Western

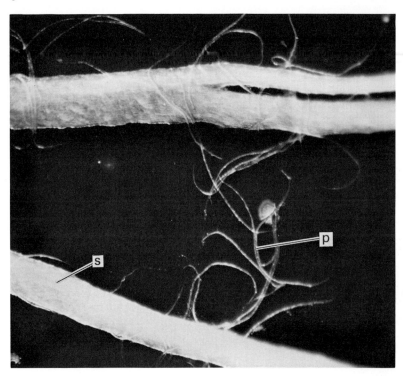

Fig. 3–4 The filiform pollen grains (p) of the sea nymph, *Amphibolis antarctica*, are transported by sea currents to the smooth, sticky stigmas (s) of the female flowers while completely submerged under the sea. (From DUCKER, S. C., PETTITT, J. M. and KNOX, R. B. (1978). *Aust. J. Bot.*, 26, 265.)

Australia in 1826. He noted the unusual filiform pollen of the sea-grass, *Amphibolis antarctica* (family Cymodoceaceae), which disperses in seawater like cottonwool in long rope-like masses in its search for the elusive female flower (Fig. 3–4), as the sexes are borne on separate plants. The remarkable property of sea-grass pollen is that it can bend and flex in sea currents, essential in grains of such length, eel grass, *Zostera* (family Zosteraceae), having pollen nearly 3 mm long, while *Amphibolis* is more than 5 mm in length. This is achieved by the virtual loss of the outer exine layer.

Nearly all these aquatic plants are monocotyledons. There is an aquatic dicotyledon with a diversity of pollination mechanisms, the water starwort, *Callitriche*. *C. verna*, flowers above water, and has pollen with a thick exine; *C. autumnalis*, in contrast, is totally submerged with pollen recorded as having no exine, but is covered with lipids perhaps to seal it from its aquatic environment on its journey to the stigma.

4 Pollen and Fertilization

Receptive stigmas may receive a variety of pollens, both from their own species and from foreign species, and are able to select out compatible pollen for seed-setting. This selection is believed to be mediated by a specific binding between macromolecules located on the stigma and in the wall sites of the pollen grain.

4.1 The two domains of the pollen wall

The walls of all pollen grains are adapted for the conveyance of proteins and other material derived both from the pollen itself and from the parental cells of the sporophyte. The two layers of the pollen wall,

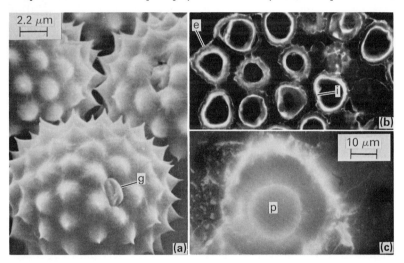

Fig. 4-1 The two domains of the pollen grain wall revealed by fluorescent antibody staining of ragweed pollen. (a) shows the appearance of whole grains by scanning electron microscopy while (b) and (c) show freeze-sectioned grains after treatment with fluorescent-labelled rabbit antibodies to pollen diffusate. In (b), antibody binding to the exine proteins (e) and intine proteins (i) is evident as white areas in the grains which were sectioned within the anthers to avoid diffusion artifacts. In (c), mature dry pollen was freeze-sectioned directly in gelatine medium after exposure to the medium for only 30 sec. In this short time, the proteins have diffused from their wall sites and are seen as a halo around the grain (p). ((a) is from HOWLETT, B. J., KNOX, R. B. and HESLOP-HARRISON, J. (1973). *J. Cell. Sci.*, **13**, 603–19; (c) is from KNOX, R. B. and HESLOP-HARRISON, J. (1971). *Cytobios*, **4**, 49–54.)

exine and intine, represent two different domains in respect to the mobile constituents (Fig. 4–1). The cavities in the exine, crypt-like in tectate types, and open in the net-like pilate types, carry proteins and glycoproteins secreted by the anther tapetum, often sealed within an oily surface layer. In contrast, the intine layer during deposition receives proteins from the young spore and later from the vegetative cell of the pollen grain. These are stored within the polysaccharide matrix of the wall in the form of tubules or leaflets. In aperturate grains, the intine storage sites are concentrated beneath the germinal apertures; while in non-aperturate types, the proteins are held in the intine all around the grain.

These products, of different genetic origin, are held within the pollen wall and are rapidly released when the grain is moistened, for example when it is received at the receptive stigma surface. The routes and kinetics of release of the wall-held proteins can be studied by simple experiments (see Chapter 7). First to be released are the exine-held proteins, which diffuse from the wall cavities within seconds of moistening (Fig. 4–1). The intine-held proteins follow later, within 2–3 minutes, diffusing from the germinal apertures in pollens where the intine is sealed efficiently by the nexine, or from all around the grain where the nexine is perforated by channels. To date, most studies of the pollen wall proteins have been histochemical or have involved qualitative electrophoretic examination of the proteins in pollen extracts. These have shown that the pollen wall sites contain a number of enzymes, antigens, and glycoproteins. One total quantitative analysis of the high molecular weight material in a pollen extract shows that protein and carbohydrate are present in approximately equal proportions with lipid as a minor component.

Plant lectins or agglutinins extracted from the seeds of legumes, which can act as 'plant antibodies', bind with remarkable specificity to certain plant glycoproteins. Lectins are basically carbohydrate-binding proteins, and they were first detected by their ability to bind to the glycoprotein coats of animal cells, causing agglutination or clumping of the cells. This property can be specifically inhibited by the prior addition to the lectin of a simple sugar. The lectins from different species of legume seeds have different sugar-binding requirements. The reason that the lectins are able to bind to glycoproteins at cell surfaces is that they recognize and bind to the terminal sugar in the chain attached to the protein. The lectin from the jack bean, *Canavalia ensiformis*, is a protein with the ability to interact with glucose and mannose, but not with galactose. It has been isolated and purified, and called Concanavalin A. It will form precipitates in double diffusion tests in agar gel with many pollen extracts, indicating that glycoproteins with the requisite sugar specificity are present in the pollen wall sites. Lectins can also be labelled with a tracer, and when Con A is labelled with fluorescein isothiocyanate, it binds to intine and exine protein sites in the pollen grain, demonstrating that glycoproteins are present in pollen walls.

4.2 Male–female recognition

Success in pollination requires a very fine degree of reciprocity and co-adaptation between the pollen and the stigma, the portal of entry to the female egg cell. After alighting on the stigma, the pollen grain may germinate, producing a pollen tube that transports the two sperm cells to the embryo sac where they are released. One fuses with the female

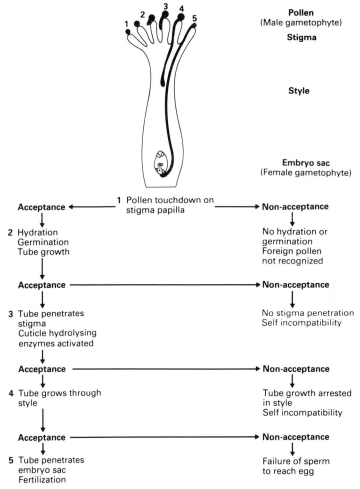

Fig. 4–2 Diagram representing the behaviour of pollen in terms of the events leading to compatible pollination and the various incompatibility options. The hand-like structure represents the pistil, and the finger-like processes represent individual papillar cells. The events are based on the observations of pollination in the Iridaceae, Compositae and Cruciferae referred to in the text.

gamete, the egg; the other with the primary endosperm nucleus to produce the embryo and endosperm in the double fertilization event characteristic of flowering plants (Fig. 4-2). Mismatch at any point during the sequential events produces an effective barrier to delivery of the male gametes.

A variety of pollens may make contact with the stigma surface, either carried there by air currents (Fig. 4-3) or by pollinating animals. The stigma of *Gladiolus* has a dry but sticky surface. In this stigma, experiments have shown that pollen from botanical families other than the Iridaceae to which *Gladiolus* belongs, is usually ignored, and neither swells nor germinates (Fig. 4-4). Pollen from within the Iridaceae usually hydrates.

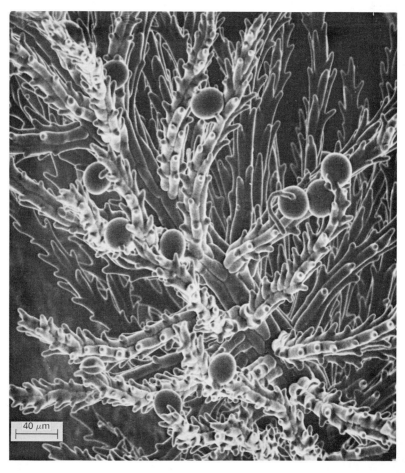

Fig. 4-3 Germination of grass pollen on stigma surface of ryegrass.

This is followed by an immediate release of exine proteins through the surface micropores, followed within minutes by release of intine proteins. These events signal the initiation of pollen-tube growth and germination. At this point, a second recognition event occurs: the activation of enzyme systems which enable the pollen tube to digest its way through the wall layers of the stigma and enter the stigma cell. The pollen tubes grow between rather than through the stylar cells, or in *Gladiolus* and *Lilium* through the style canal which is filled with a mucilage which supports pollen tube growth.

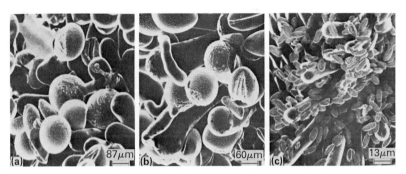

Fig. 4–4 Scanning electron micrographs of living *Gladiolus* stigma, 2 hours after pollination, with (a) compatible *Gladiolus* pollen, (b) *Crocosmia aurea* pollen and (c) *Gloriosa rothschildiana* pollen. (From KNOX, R. B., CLARKE, A. E., HARRISON, S., SMITH, P. and MARCHALONIS, J. J. (1976). *Proc. Nat. Acad. Sci. (U.S.A.)*, **73**, 2788.)

Even at this late stage of pollination, completion of an incompatible mating can be stopped (Fig. 4–2, steps 4 and 5). Some forms of self-incompatibility involve pollen tube arrest at this stage. The final stages of pollen-tube growth are penetration of the embryo sac, when the tube enters a support cell adjacent to the egg, and by means of a pore which opens in the tube, discharges the two gametes, probably with some explosive force. One gamete unites with the egg nucleus to form the zygote, the other with the primary endosperm nucleus to trigger formation of endosperm. Even at this late stage, cut-off points exist to prevent fertilization, including failure of the pollen tube to release the gametes, or if released their inability to bind to the egg cell. Both mechanisms would prevent fusion, and incompatibility at this stage operates in cocoa, *Theobroma cacao*, and day lily, *Hemerocallis*, following selfing.

4.3 Control of pollination

Active rejection responses following incompatible matings were first noted by Charles Darwin in 1882. He observed that the pollen masses in certain orchids turned brown and necrotic after selfing instead of separating and germinating. More recently, the polysaccharide callose has been implicated in the rejection response in both germinating pollen and in some cases, also in stigma cells in contact with rejected pollen (Fig. 4–5). The nature and rapidity of these visible manifestations of incompatibility are important consequences of the recognition reactions between pollen and stigma.

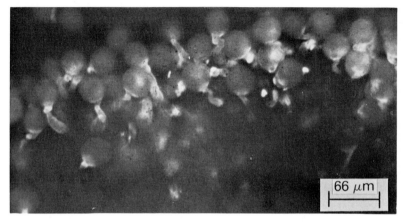

Fig. 4–5 Fluorescent micrograph showing callose produced in pollen tubes in response to incompatible mating in *Iberis* (Cruciferae).

The interacting cell surfaces of pollen and stigma can be manipulated both physically and chemically to alter their capacity for mutual recognition. For many years, I. V. Michurin and his successors in the U.S.S.R. showed that wide incompatibility – between different genera and species – could be overcome by applying mixtures of pollens to receptive stigmas. In 1968, R. F. Stettler in Seattle, working with incompatible species of poplar trees, was able to disguise normally incompatible pollen by mixing it with compatible pollen sterilized by γ-irradiation, and obtained hybrid progeny. R. B. Knox, R. R. Willing and A. E. Ashford carried out further experiments with incompatible black and white poplar species in Canberra. They found that the incompatible species pollen could equally successfully be disguised with diffusates of compatible pollen to obtain hybrid progeny. An implication of these experiments is that the pollen-wall proteins are vital for species recognition on the stigma surface. In subsequent experiments with poplar pollen, and with various crucifers, solvent washing of pollen also changed its recognition potential, demonstrating that the integrity of the natural interacting surfaces is also essential for the expression of the recognition response.

5 Aerobiology of Pollen

Pollen constitutes but a small part of the aeroplankton or air spora present in the atmosphere. The most frequent particles of biological origin are microorganisms, especially the spores of fungi. For example, pollen represents only 2% of the air spora detected annually in Cardiff. The others are fungal spores belonging to various groups: Fungi imperfecti 43%; Basidiomycetes 37%; Ascomycetes 17%; and Phycomycetes less than 1%. Particles, when dispersed in air, are termed *aerosols*. The presence of bacteria and viruses in aerosols is less easy to detect, but algae, leaf hairs, seeds, plant fragments, and volatile materials including scents and terpenes also occur. The terpenes are oily substances released from the leaves of trees in sunlight, and may form a blue haze in the atmosphere, or aggregate and polymerize in sunlight, forming a brown-black air soot. The atmosphere may also contain other particulates including bushfire ash, industrial ash spheres and cenospheres from incomplete fuel combustion. Aerobiology is concerned with the behaviour of a suspension of particles, both viable and non-viable, whose transfer from one site to another is governed by atmospheric properties and these aerosols may travel short distances, or may be blown into the upper strata of the atmosphere and travel long distances before they are deposited.

5.1 Pollen transport in the atmosphere

The fate of particles such as pollen in an aerosol during dispersal in the atmosphere was pictured by J. M. Hirst in 1973: 'to anything as small and light . . . air is a fairly viscous medium. If you and I were immersed in treacle we might share a falling speed of only a few mm or cm/sec., be unable to be projected far through it and be involuntary passengers of its motions.' The atmosphere has been well-described as a restless ocean of air. It is divided into a number of zones: the troposphere, stratosphere, ozonosphere and mesosphere, named in order of distance from the earth's surface. The troposphere houses nearly all the air necessary for life. It is said to be in convective equilibrium, in contrast with the outer layers which are in radiative equilibrium. Energy from the sun heats the atmosphere mainly by convection, and air temperature decreases with increasing altitude within the troposphere.

Close to the surface of the earth is an extremely thin layer of air bound by molecular forces, which has above it a film of moving air known as the *laminar boundary layer*, in which air flow is parallel to the surface. Under windy and turbulent conditions, this layer may be only a fraction of 1 mm thick, while on a clear calm night, it may be several metres deep. The layer acts like a shield preventing dust particles from reaching the turbulent wind eddies above. Once launched into the atmosphere, however, particles such as pollen grains are subject to forces resulting in both vertical and horizontal motion, and their fate is determined by the stability of the atmosphere.

When compared with a point source such as a factory chimney, most biological sources of particulates are both weak and dispersed. Single point sources of biological aerosols have been created experimentally to study the dispersal of particles such as pollen grains. Most biological sources are likely to be area or line sources. Despite their shape at source, their fate is much the same, and most pollens are redeposited on the ground within a few metres. However, a variable proportion, termed by P. H. Gregory in 1962 the *escape fraction* is able to diffuse upwards and travel longer distances before it is eventually deposited. Most airborne pollen grains are about 25 μm in diameter and, obeying Stoke's law, would have a free-fall velocity of 7.5 cm sec^{-1}. If released from a height of 1 m, assuming a wind speed of 10 knots, the pollen would travel 67 m before deposition; while if released at a height of 20 m, it would travel 1333 m. However, because of other factors, such as turbulence, many pollens are deposited close to their source.

Vertical movements give rise to the mixing of parcels of air containing dense aerosols with lower density aerosols from higher levels (Fig. 5–1). When the particles reach their new position, they are transported in the mean direction of the wind, and at the same time tend to be dispersed vertically under unstable conditions, and most are deposited on the

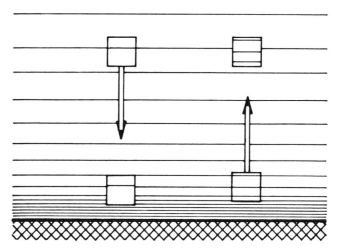

Fig. 5–1 Mixing of aerosols containing pollen within parcels of air as a result of vertical movements. (From NILSSEN. 1973.)

ground a short distance from take off (Fig. 5–2). Their fate is different under stable atmospheric conditions, when they may travel long distances without deposition. The conditions ideal for long distance transport are, first, unstable air conditions for the pollen to be propelled upwards, and then stable conditions for prolonged transport.

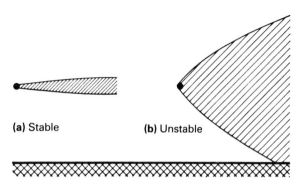

Fig. 5–2 Diffusion of aerosols containing pollen in (a) stable and (b) unstable situations. (From NILSSEN, 1973.)

In the troposphere, there is often a diurnal variation in stability. The layers close to the surface tend to be stable during the night because of the radiative cooling of the earth's surface. By mid-morning, however, the heating of the earth's surface by the sun's radiation leads to instability. Between 10.00 and 12.00 hours daily, pollen grains released under such conditions may reach high altitudes in the turbulent atmosphere. In contrast, at night, when low-temperature inversions occur, they are accompanied by a stable atmosphere close to the surface, reducing the possibility of exchange of pollen from lower and higher strata. This is brought about by the radiative cooling of the earth's surface which results in an inversion, in which temperature increases with increasing height, in a still and stable atmosphere. Such conditions favour long distance transport and pollen may spend several days in the atmosphere.

Rain or drizzle in spring and early summer during the pollen season is most effective in cleaning the atmosphere of large particles such as pollens. It appears to cause deposition of particles in two ways: particles may act as nuclei of condensation for cloud droplets and so be deposited directly encased in rain water drops, or more commonly, deposition may occur from washout with preformed raindrops.

Long distance transport can also occur during heavy showers and storms. These develop during very unstable conditions, with regular vertical air movement (Fig. 5–3). If the showers cover great distances, as along a cold front, the particles will be similarly dispersed. A proportion of the particles will be caught in downward currents and deposited within the storm area. In central and eastern Europe, snow or dust storms occur in winter, in which light humus soil (loess) and particles such as pollen are carried long distances in the troposphere, to fall as *red snow* or rain.

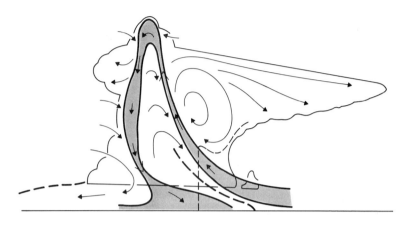

Fig. 5-3 Schematic diagram of behaviour of air currents containing pollen aerosols (hatched) during a well-developed thunderstorm. (From NILSSEN, 1973.)

The first demonstration of long distance pollen dispersal was by the Swedish botanist, H. Hesselman, in 1919 who placed adhesive plates on two lightships moored in the Gulf of Bothnia. The nearest ship was 30 km offshore, and collected 100 000 pollen grains during a period of over one month in early summer. The other ship was 55 km from the shore, and collected just over half as much pollen as the other, mainly pollens of birch, pine and spruce. An even more spectacular demonstration was later given by G. Erdtman, the famous Swedish palynologist, who devised a vacuum-operated suction trap to collect pollens in measured volumes of air. In 1937, he went on a voyage across the Atlantic Ocean during the pollen season, and succeeded in monitoring significant quantities of alder, birch, pine, oak, sedge and grass pollen in mid-ocean.

Various methods are now available for monitoring pollen in the atmosphere. The simplest method, using gravity-slides, involves the exposure of slides coated with an adhesive such as double-sided sellotape or 'vaseline' petroleum jelly to the atmosphere for a period of time, usually one day. The particles dropping by gravity onto the slide are then scored. These methods give a qualitative picture of the different kinds of pollen, but the results may bear little relationship to the numbers present in the surrounding air. To gain a more accurate impression of the quantitative relationships among atmospheric particles, volumetric spore traps must be used. In these, measured volumes of air flow over an adhesive surface on which the particles may be impacted in proportion to their occurrence in the air. J. M. Hirst in 1952 developed a spore trap which provided an estimate of atmospheric particles within the size range 3–40 μm (fungal spores and most pollens) with an efficiency varying between 55 and 90%. In the Hirst trap, air to be monitored is drawn into an orifice which points into the prevailing wind. The particles impact on an adhesive slide or tape that moves past the orifice at a constant rate, usually 2 mm per hour. This trap has the advantage that variation in pollen concentration with time can be measured, and when a tape is used it can be left unattended for seven days. The rotorod, a rotating impaction trap, is a simpler device which permits volumetric sampling with high efficiency. It has two U- or H-shaped antennae which provide the adhesive collection surface. These rotate around drawing air past at a known volume with time. The particles impact on the forward sides of revolving square metal rods or the edge of a microscope slide as the collecting surface. The rotorod can be operated from a battery and is most useful for operation for short periods. Collecting efficiency varies according to the size and density of the particles.

5.2 The pollen calendar

Tree grass and weed pollens are common in the atmosphere in most climates. While certain weeds are of almost ubiquitous occurrence, most pollens of temperate climates will be quite different from their tropical counterparts. In tropical areas, plants tend to flower for longer periods, so that pollens are likely to be present in the atmosphere for most months of the year. The pollen season in temperate climates is restricted to the warmer months of the year from late winter through to autumn. As the flowering season progresses, different pollens are present in the atmosphere, so that each type displays a particular seasonal periodicity (Fig. 5–4). In Europe, North America and Australia, where aerobiological records are kept, the seasonal progression involves first tree pollens in winter and early spring. The pollens of birch, alder, hazel, oak, ash and elm lead the pollen calendar (Fig. 5–4). In late spring and early summer, the grass season begins, followed closely by various weeds; for example, nettle, dock, sorrel, plantain, and in North America by various amaranths and ragweeds in the autumn.

The pollen content of city air from London has been compared with that of a country sampling station at Rothamsted in Hertfordshire. Unexpectedly, the total pollen rain monitored over the city was actually greater than that at the country site, largely because of the high levels of plane tree pollen in the city, where it is a common street tree. For other tree pollens, Rothamsted showed higher levels, and similarly with grass pollen. The weed pollens, however, had greater concentrations in the city, especially plantain and nettle.

In Germany, the seasonal incidence of grass pollen is about four times as long as for birch pollen. Both the time of commencement and duration

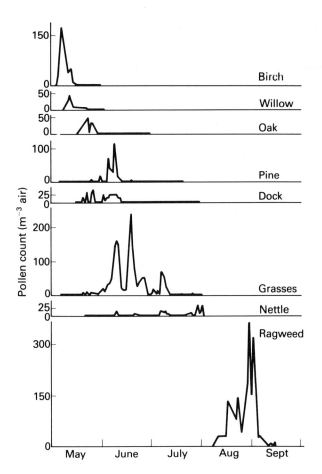

Fig. 5-4 Idealized diagram of a typical pollen calendar showing pollen counts of trees, grasses and weeds common in cool temperate climates. (Data for all plants except ragweed based on Utrecht, Holland, from data of van den Assem, in NILSSEN, 1973; ragweed data for Brookhaven, New York, from OGDEN *et al.*, 1974.)

of the grass season has been shown by Stix to be dependent on elevation above sea level and geographic position. In a survey of pollen incidence at eight centres, the shortest and latest seasons were found to be coincident – occurring at the most northerly and highest altitude stations. However, marked differences in onset of the pollen season occur from year to year. Hazel pollen was released in early February in one year, but not until the end of March in another year. Birch pollen, which appears later, is regularly released from mid to late April. Likewise, the total amount of pollen released varies widely from year to year.

In atmospheric pollen counts, the different kinds of grasses contributing to the pollen count cannot be identified, though it has proved possible to distinguish cereal from wild grass pollens. Also, each grass species has a characteristic flowering period, and as the grass flowering season progresses, different genera come into flower, maintaining the atmospheric pollen levels (Fig. 5–5).

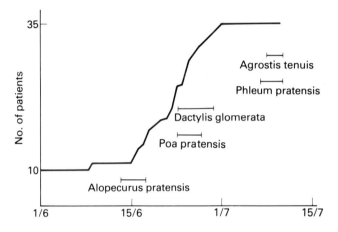

Fig. 5–5 Seasonal debut for 35 Swedish hay fever patients compared with the pollination periods of various grasses. (From NILSSEN, 1973.)

5.3 Circadian rhythms of pollen emission

Anthers of different flowers dehisce at particular times of the day, so the concentrations of airborne pollen in the atmosphere might be expected to increase about the times of emission. Daily patterns of ragweed pollen incidence at the height of the season in August and September have been monitored using volumetric traps. The peak of pollen release occurs in the morning, at any time from 06.00 till 14.00 hours depending on the weather conditions. Time of release is dependent on both pollen dehiscence from the anthers and its emission into the atmosphere, which is not as straightforward as it might appear. Ragweed flowering, including filament extension and anther dehiscence, is induced primarily by changes in temperature and relative humidity, such as occur commonly in the field at sunrise. The pollen is not immediately dispersed, but is released in clumps of grains which are deposited on adjacent foliage and stems from which they are later blown off by the wind after drying. These two distinct phases are regulated by different atmospheric conditions:

(i) release from the anthers – governed by time of day, temperature and relative humidity — followed by partial deposition on adjacent plant surfaces;

(ii) emission into the atmosphere from plant surfaces — controlled by wind speed and turbulence.

Pollen emission, on a daily basis, may differ substantially from the mean pattern over a pollen season. E. C. Ogden and co-workers at the Brookhaven National Laboratory observed the importance of weather conditions on ragweed pollen emission, which began only in the unstable, drier air following sunrise, and continued until rain fell or the wind became calm (Fig. 5–6). Pollen detected after the morning peak is considered to be that blown from plant surfaces, emission terminating when no more pollen can be removed from the foliage.

Unlike ragweed, grass pollen shows little tendency to clump together on release, being fine, smooth and dry. Nevertheless, much of the pollen in dense strands of grasses may be deposited on the surfaces of nearby leaves and stems like a yellow rain. A number of grasses, including Yorkshire fog (*Holcus lanatus*), fescue (*Festuca rubra*), and cocksfoot (*Dactylis glomerata*), show both morning and afternoon peaks of pollen liberation in the field.

In the city atmosphere of London, the peak of atmospheric grass pollen occurred between 15.00 and 20.00 hours daily. There is a slight delay of 1–2 hours between the peak at a country sampling site, Rothamsted, and the peak in London. In Holland, over the city of Utrecht, the peaks of grass pollen occurred between 15.00 and 18.00 hours. On some days, however, three peaks were detected, at 05.00, 11.00 and 17.00 hours, at the height of the flowering season. In contrast, grass pollen counts in the city atmosphere of Melbourne, Australia are greatest at night, probably because of low temperature inversions which tend to concentrate pollen aerosols.

The flowering of timothy has been investigated in North America. In Ohio, M. W. Evans in 1916 observed that flowering occurred during the night, commencing at 22.00 hours, reaching a peak at 02.00 and ceasing by 06.00 hours. M. D. Jones, in Oklahoma in 1952, found that release occurred from 04.00 to 09.00 hours. More definitive information has been obtained by experiments with Hirst spore traps sited just above foliage level in the centre of a timothy field at Upton, New York. Ogden and co-workers detected the first release of pollen into the atmosphere before sunrise, but the peak of pollen was found on average about three hours after sunrise. Emission began any time from 21.00 to 06.00 hours. Figure 5–6 shows pollen emission over two days; 'during the early morning of 10 July, humidity was relatively low, the air was stable for only a brief period, and emission became heavy at 02.00 in spite of very low wind speeds. A peak was reached at sunrise and emission was mostly ended by 08.00. On the following day the air remained stable, the humidity high, and the winds low until after sunrise. No significant emission took place until 05.00. A sharp peak occurred from 06.00–07.00

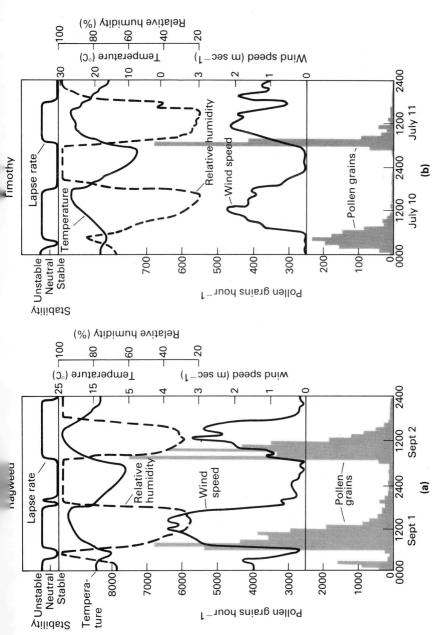

Fig. 5-6 Variation of hourly pollen counts with atmospheric conditions. (a) Ragweed; (b) Timothy. (From OGDEN, E. C., HAYES, J. V. and RAYNER, G. S. (1969). *Am. J. Bot.*, **56**, 16–21.)

as the wind speed increased, the humidity dropped, and the air became unstable.'

The peaks of grass pollen correlated well with increases in temperature, are negatively correlated with rainfall and relative humidity and there was much less pollen in the atmosphere when the sky was half to fully clouded over. There are likely to be many airborne pollens in the atmosphere on days with fine clear weather. Rainfall will deplete the pollen in the air, but when the sun shines afterwards, rapid increases in pollen counts have been observed. Areas of high atmospheric pressure are favourable for pollen emission, while under low pressure or passage of a depression conditions are unfavourable. Peaks of pollen in the atmosphere are associated with high temperatures.

In general, the pollens of trees and weeds show a similar diurnal periodicity to those described for ragweed and grasses. Typical of tree pollens is that of birch, and in Darmstadt, Germany, the maximum count occurred each day between 12.00 and 18.00 hours during the season.

A major weed in Europe is the nettle, *Urtica*. At Utrecht, in Holland, A. van den Assem in 1973 showed that the diurnal peaks in nettle pollen are similar to those for grass pollen, except that its main occurrence is towards the end of the grass pollen season (Fig. 5-4). On favourable days, the nettle pollen concentration rose dramatically, while the grass pollen concentration showed a reduced peak. This has been called the *depletion effect*, since towards the end of the pollen season for a species, there is no 'bank' of mature flowers ready to open when favourable conditions arise, on warm, dry sunny days.

The value of pollen counts, expressed on an average daily basis, which appear each day in the press in many cities, is reduced because of the marked variation in pollen concentration in the atmosphere during the day. In addition, pollen concentrations are generally highest at ground level, reducing exponentially with height above ground. However, in the concrete canyons of cities this may not be the case. R. R. Davies has recorded in London that pollen concentrations at street level were actually lower than on the top of buildings. The buildings may shelter pockets of air from the pollen-laden airstreams blowing over the city, and on sunny days the warm air rises from the streets and discourages deposition of pollens. North American aerobiologists have attempted to predict the onset and magnitude of the ragweed pollen season from analyses of pollen count in relation to weather parameters of previous seasons; this has proved feasible if enough quantitative data for a number of seasons is available. In those areas where grass pollen is a problem in spring and early summer, the possibilities of forecasting are decreased by the sheer length of the pollen season, often extending over 12–14 weeks and by the variety of grass genera that contribute to it.

6 Pollen and Man

6.1 Role of pollen in allergic disease

Environmental agents, such as pollen, may initiate the allergic response in susceptible humans. People who are allergic have an altered capacity to react to potential allergens or are said to be hypersensitive to them. Pollens have been implicated in several allergic diseases, including asthma (allergic lung disease), hay fever (allergic rhinitis), together with several eye, skin and respiratory disorders. A characteristic feature of pollen sensitivity is its seasonal pattern of occurrence, usually at the time when the pollen is most frequent in the atmosphere.

The kinds of particles present in the atmosphere that are allergenic vary according to the climate, geography and vegetation, and little information is available on their relative importance. In Sweden, S. G. O. Johansson recently carried out a survey of more than 300 allergic patients, and demonstrated that most had reactions to animal danders, that is, fur, hair or skin (Fig. 6–1). Some 18% were sensitized to horse, 17% to cat and 5% each to cow and dog dander. Of the remainder, 11% responded to tests with house dust, 16% to food, including milk, eggs and fish, and 30% to pollen. Since much of Sweden is covered by birch forest, it is not surprising that 16% of patients were sensitized to birch pollen, while an almost equal number were allergic to grass pollen. In North America, ragweed pollen replaces birch in importance, while in temperate Europe,

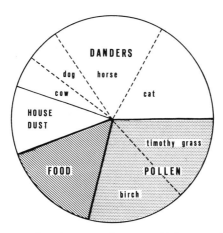

Fig. 6–1 Frequency of environmental agents which cause allergy in patients. (Modified from JOHANSSON, S. G. O. in EVANS, R. (ed.) (1975). *Advances in Diagnosis of Allergy*: RAST. 7–16. Symposia Specialists, Miami, U.S.A.)

southern Australia and New Zealand grass pollen is the major allergenic pollen, and tree pollen is much less important.

One of the first studies that related pollen present in the atmosphere to allergic disease was carried out by Andrup in Norway between 1939 and 1941. He monitored pollens in the atmosphere using the vacuum spore trap which had been developed in the 1930s by Erdtman (see p. 44), and found that birch pollen predominated in May and early June, and grass pollen in late June and July. When the symptom scores of hay fever patients were compared with pollen incidence (Fig. 6-2), peaks in the pollen count coincided with increased intensity of hay fever in the group of patients. O. Andrup was able to compare the seasonal 'debut' of symptoms in a group of 35 hay fever patients with the pollination periods of the grasses (see Fig. 5-5). The period when most patients were affected corresponded with the flowering season of cocksfoot and timothy grasses, well-known sources of allergenic pollen used for clinical tests.

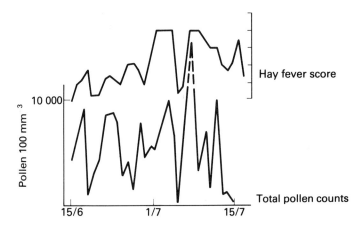

Fig. 6-2 Hay fever intensity compared with atmospheric pollen counts in Oslo, Norway. (After ANDRUP, O. (1945). *Norsk. Videnskab. Akad. I. Nat.-Naturv. kl.* no. 5; and HAVNEN, J. in NILSSEN, 1973.)

Comparisons of the grass pollen count in London with patient's hay fever symptoms also revealed a coincidence of symptom scores and pollen counts. A. Frankland and R. R. Davies in London noted that when grass pollen concentrations in the city atmosphere rose above 50 grains m^{-3}, all patients in the area who were clinically sensitive to grass pollen experienced symptoms. In Australia, symptom scores of a group of 19 Melbourne schoolchildren with childhood asthma and laboratory-tested lung sensitivity to ryegrass pollen showed marked increase in symptoms during the grass pollen season in spring and early summer. Birch pollen counts in Sweden have been related to clinical symptoms of allergic eye disease in children. Autumn hay fever symptoms in patients allergic to ragweed pollen in Minnesota, U.S.A., were generally coincident with high pollen counts.

The reasons why pollens cause asthma in some persons and hay fever in others are obscure. Hay fever is induced by pollen when it makes contact with the upper respiratory tract; the nostrils, oral cavity (mouth) and eyes. A hay fever patient may suffer considerable irritation of the eye immediately following contact of a pollen grain with its surface, which is likely since grass pollens are large enough to be impacted on the eye at relatively low wind speeds. In contrast, asthma is a disease of the lungs. If inhaled, pollens may be deposited in the uppermost ciliated portion of the respiratory tract. During an asthmatic reaction, symptoms develop in the deeper, non-ciliated parts of the lung, which show accumulation of fluid and secretions in the terminal bronchioles. Symptoms may appear immediately following pollen exposure or be delayed for some hours. Entry of inhaled pollen to the lungs is regulated in several ways. It depends on the size of the pollen grains and the diameter of the airways. From the trachea to the terminal bronchi, the airways divide into two equal branches about 19 times, and their cross-sectional area is reduced from 13 mm^2 to 3 mm^2 with a consequent reduction in air flow rate from 150 ml s^{-1} to about 1.5 ml s^{-1} in the terminal bronchioli. Particles as large as pollen cannot reach the lungs in the inhaled aerosol, the nasal cavity filtering them out by inducing a high degree of turbulence in the airflow. Inhaled particles with a diameter greater than 30 μm, which includes all pollens and most fungal spores, are deposited in the trachea and upper bronchi.

There are two ways in which pollens and other particles are eliminated from the respiratory tract. First, the surface of the bronchi and bronchioli is coated with ciliated epithelial cells. These beat in the direction of the nasal cavity at a frequency of 1300 min^{-1} in a healthy person. A mucous layer lies on top of the cilia, and moves in the direction of the trachea with a speed of 1.5 mm min^{-1}, so that the nasal-pharyngeal cavity is reached within an hour. Infections and irritants reduce the rate of movement, but this can be increased by coughing. Any particles deposited in the lungs are removed by alveolar fluid and by macrophages, white blood cells, which can partly break down the inhaled material and drain it into the terminal bronchioli or regional lymph nodes.

Proof that grass pollen is unlikely to gain direct entry to the lungs has come from experiments in which pollen of June or Kentucky blue grass, *Poa pratensis*, was labelled with radioactive technetium, ^{99m}Tc. When inhaled, the pollen was deposited in the oropharynx and did not reach the trachea, but was swallowed and most of the pollen accumulated in the stomach. Here, a process called *persorption* may take place. Pollen, taken orally, passes through the stomach, but a proportion passes directly from the stomach into the bloodstream within 45 minutes. The significance of persorption in the initiation of allergic reactions remains to be assessed.

6.2 Allergens in pollen

What is an allergen? It is not the pollen grain itself, but factors located on or within it, that may induce allergic disease. Allergens are proteins or glycoproteins that are capable of eliciting the formation in susceptible humans, of specific skin-sensitizing or reaginic antibodies through the body's immune system.

Before we can understand how an allergen works, we need to understand the nature of the allergic response. It was first defined by C. von Pirquet in 1906 as *'the acquired, specific, altered capacity to react'*. The response is obviously bound up with the immune system, its ability to discriminate between self and non-self and take defensive action. J. Pepys, a distinguished allergist and immunologist at the Brompton Hospital in London, has shown that each part of von Pirquet's definition has important implications. *Acquired* means that there must have been previous exposure to the allergen to stimulate the immune system and develop hypersensitivity. It also means that once identified, steps can be taken to avoid unnecessary exposure to the allergen. *Specific* refers to the very precise molecular relationship that exists between the allergen and the corresponding antibodies produced in response. The surface determinants of both molecules carry its chemical message. Related allergens may carry some common determinants, so allowing for a degree of cross-reaction between them. *Altered capacity to react* describes the different response induced by the same allergen after antibodies have been produced in the body against it. The allergic response may be increased, as in hypersensitivity, or it may be decreased as a result of increased immunity.

Two types of allergic response to pollen have been described: *type 1*, or immediate hypersensitivity, is characterized by the rapid appearance of skin weals at the site of allergen contact; it is demonstrated by the skin or prick test, and is considered to be mediated by skin-sensitizing or reaginic antibodies, termed immunoglobulin E (IgE); and *type 3*, or delayed hypersensitivity, appears several hours after allergen contact; it is mediated by precipitating antibodies, immunoglobulin G (IgG) produced by the body after challenge by foreign antigens.

Type 1 or immediate hypersensitivity covers the majority of pollen-induced allergic responses. The first evidence that the reactions to an allergen are mediated by a specific factor (now known to be IgE) was obtained more than 50 years ago by the famous medical experimenters C. Prausnitz and H. Küstner. Küstner was allergic to fish, and developed a weal reaction to skin tests with cooked fish extract. Prausnitz was not allergic to fish, and showed no reaction to tests with fish extract. In the experiment, serum from the allergic Küstner was transferred to Prausnitz by injecting subcutaneously. One day later, fish extract was applied to the area of the injection, and a positive weal reaction was elicited immediately, demonstrating that Küstner's serum had contained a specific factor promoting the allergic reaction to fish allergen. The test works equally well for pollen.

The specific factor, IgE, was identified in 1966 by K. and T. Ishizaka and co-workers. They have shown that it is made in antibody-forming cells of lymphoid tissues of allergic individuals in response to initial exposure to allergens such as pollen. The IgE produced circulates in the serum in the bloodstream, and owes its biological importance to its affinity for certain epithelial and mucosal cells, *basophil* and *mast cells*, to which it becomes attached by its footpiece, the F_c region of the molecule (Fig. 6-3). IgE molecules may remain bound to mast cells for periods of several weeks, each cell having up to 100 000 IgE molecules on its surface. Each molecule has two arms with a terminal recognition site for its specific allergen and is in communication with the mast cell membrane through the membrane glycoprotein to which it is attached.

Later, when the allergen molecules are again encountered, they bind to pairs of adjacent IgE molecules on the mast cell surface. This binding interaction triggers the rapid release of tissue mediators, pharmacologically active substances from granules secreted by the mast cell (Fig. 6-3). These include histamine and enzymes which bring about the symptoms of allergic disease. The reaction consists of itching, inflammation and wealing, and is accompanied by eosinophil cell infiltration at the site of testing. It can readily be inhibited by therapeutic drugs if they are administered prior to challenge. Similar symptoms are evident in nasal, eye and bronchial responses to allergen, including wheezing, coughing and tightness of the chest.

What proportion of the population develop allergic disease from exposure to aeroallergens such as pollen? This is a difficult question to answer because of the paucity of reliable statistics. In the U.S.A., some 15% of the population are said to be affected by hay fever from ragweed pollen. In Britain, hay fever is said to afflict 3% of the community, and in Australia, a study of disease incidence in a country town in Victoria revealed that 11.4% of the population suffered from hay fever, and 4.6% from asthma, while in the city of Melbourne, 10% of schoolchildren had seasonal wheezing symptoms.

Pollen allergens are proteins or glycoproteins, and have been characterized for both ragweed and grass pollens. T. P. King and co-workers at the Rockefeller University in New York isolated and purified the allergens from ragweed pollen in 1964, naming the two most potent antigens E and K. They have proved to be acidic proteins comprising two sub-units, an α chain of molecular weight 21 800 and a β chain of 15 700, giving a total molecular weight of 38 000. The two chains are readily dissociated by heat, and are linked by covalent bonds. Antigen E is present in four different forms, that are all immunologically similar but differ in isoelectric points. The allergens of grass pollen are equally complex, and three groups of heat-stable glycoproteins are the principal

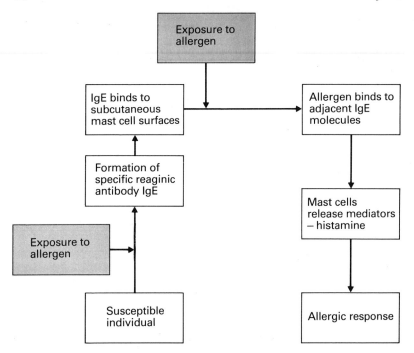

Fig. 6–3 Diagram showing sequence of events in production of allergic response.

allergens. The two principal groups of allergens, named groups I and II by D. G. Marsh and collaborators, have molecular weights of 30 000 and 10 000 and each have several isoallergens differing in isoelectric point.

Allergen purification is carried out by standard biochemical methods for protein separation, and the fractions obtained can be tested for allergenicity by several methods. Skin or prick tests, or bronchial provocation tests provide a rapid means of testing in allergic subjects, by comparing the response to known concentrations of whole pollen extract with those of the diluted fractions. Also, the ability of allergens to release histamine from sensitized leukocytes has been utilized. The affinity of IgE in the sera of allergic subjects for its specific allergen provides the basis of the radio-allergo-sorbent test (RAST). It provides a relatively simple and highly specific means of measuring the concentration of IgE in the serum of normal healthy people and in allergic individuals. Since IgE is present at much lower concentrations than IgG, probably by a factor of 500 000 times less, it cannot be detected by immunodiffusion methods. Instead, use is made of its affinity for specific allergen. The allergen is bound by biochemical methods to a solid phase, such as a filter paper disc (Fig. 6–4).

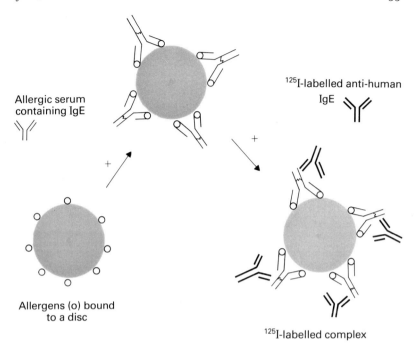

Fig. 6-4 Principles of radio-allergo-sorbent test (RAST).

For testing, the allergen-coated disc is first incubated with a drop of the patient's serum which contains the IgE to be estimated and, after several hours, it is washed to remove unbound proteins that will not interact with the allergen molecules. The specific IgE on the disc is detected by a radioactive test, in which an antiserum raised in a sheep against human IgE is labelled with the tracer ^{125}I. This reagent is applied to the disc, incubated to allow binding, then washed and the amount of radioactivity present estimated with a gamma counter. This is related to the amount of IgE on the disc, and is computed in units of IgE in relation to a standard control serum used in the test. Results show that healthy people have, in general, a low level of IgE, around 14 units ml^{-1} of serum, while in allergic patients, levels rise in excess of 100 units ml^{-1}. The test provides a high degree of discrimination between allergic and non-allergic people, and has been in laboratory use since 1969.

IgE antibody concentrations in serum are not constant, and will increase with challenge by an allergen. In hay fever patients, S. G. O. Johansson in Sweden has found that the lowest annual levels occur in the two weeks before onset of the pollen season. During the pollen season itself, there can be a two or three-fold increase in IgE levels, followed by a gradual decrease back to pre-seasonal levels. Results with tree, grass and ragweed pollens as allergens with proven pollen-sensitive patients have demonstrated that RAST reactivity is in good agreement with other methods of testing, such as skin tests and *in vitro* histamine release tests with leukocytes. RAST inhibition has proved a valuable test for measuring cross-reactions between pollen allergenicity of different ragweed species or different genera of grasses. The test is based on competition for binding between solid phase allergens insolubilized on a disc, and soluble allergens which may compete against them for available IgE antibodies (cf. Fig. 6–4). The soluble phase allergen acts as an inhibitor of binding to the solid phase, and as more soluble allergen complexes with IgE, less IgE is available to bind to the discs which are monitored by the RAST procedure, and will show lower activity as they lose out in the competition. G. J. Gleich and collaborators in Minnesota have been able to test the allergenicity of different extracts of the same pollen. Commercial extracts of ragweed pollen showed 1000-fold differences in potency, while grass pollen extracts were fairly similar in their IgE affinity. The allergenicity of different species of ragweed was tested by comparing their potency for IgE from ragweed patients sera with that for short ragweed, *Ambrosia elatior*. In the RAST inhibition tests, extracts of the other species were allowed to compete separately with solid phase *A. elatior* allergens bound to discs. The different ragweed species were fairly similar in their affinity, showing that patients had developed IgE antibodies to all species tested, or that the allergens of each are chemically very similar. Extracts of false ragweed, *Franseria*, had little inhibitory effect, showing only low levels of cross-reaction.

Marked differences were revealed when different grass pollens were compared. When June grass was used as solid phase, and extracts from other festucoid grasses competed for IgE from pooled sera of grass pollen-sensitive patients, three other grasses proved equally efficient in IgE-binding: meadow fescue, cocksfoot and perennial ryegrass; while two others showed low affinity: timothy and sweet vernal grass. June grass and the first three grasses are closely related, belonging to the subtribe Festuceae. The other two grasses belong to different subtribes within the festucoid grasses. Bermuda grass, from the chloridoid tribe, showed no affinity and patients in Minnesota would not have been exposed to it although in south-western U.S.A. it is an important allergen.

6.3 Role of pollen allergens in nature

The allergens of ragweed and grass pollen are located in extracellular wall sites, the intine and exine. This has been established using fluorescent antibody methods, taking considerable care to prevent diffusion from the original sites during processing by use of freeze-substitution methods, since allergens diffuse from their wall sites within seconds of moistening. Pollen diffusates contain a considerable number of protein fractions; some from the exine cavities originated from parental tapetal cells so that they are likely to carry diploid specified determinants; others from the inner intine have originated from the pollen protoplast and hence carry haploid specified determinants. This raises the possibility that different forms of the allergens may be present in the two sites; however to date, no biochemical heterogeneity has been demonstrated. Pollen-wall proteins, of which the allergens constitute some 10% in ragweed, are concerned with the initial events of pollination. The role of allergens in this process has still to be determined. The allergens are characterized in terms of their role as informational molecules, able to trigger the formation of specific IgE in man, and to bind specifically to it on the surface of human mast cells. It is not unlikely that allergen molecules, with such high information content, may perform an analogous recognition role in pollen-stigma interactions.

No tests have yet been carried out to determine whether within populations of allergenic species, such as the short ragweed, *Ambrosia elatior*, or ryegrass, *Lolium perenne*, there exist any individuals which have pollen with low allergenicity. This assumes, of course, that the allergens do not play a vital role in pollination. Nevertheless, the search may prove worthwhile as it could lead to the introduction into agricultural practice of strains of grass with pollen of low allergenicity and thus to a significant improvement in the environment for those allergic to the pollens of agricultural grasses.

7 Experiments with Pollen

7.1 Making a reference collection of pollen types

Place a drop of Calberla's fluid on a microscope slide, dissect out pollen from anther or dust pollen into stain, cover with coverglass. Store flat in dark place. Pollen exines are stained red. Calberla's fluid is prepared by mixing: 5 ml glycerol, 10 ml of 95% ethanol or methylated spirit, 15 ml distilled water, 3 drops of saturated solution of basic fuchsin in water, and 3 drops of melted glycerin jelly. Label slide with plant name, date and location. For further information see OGDEN, E. C. *et al.* (1973), *Am. J. Bot.*, **56**, 16–21.

7.2 Pollen viability

This method makes use of the fluorochromatic reaction, and requires the reagent fluorescein diacetate, and access to a fluorescence microscope equipped with blue exciter filters (BG 12). Fluorescein diacetate is non-fluorescent and is readily taken up by cells, where it is broken down by enzymes to release fluorescein, a highly fluorescent compound, which accumulates within the cytoplasm in fertile grains, and leaks out into the medium in sterile grains. For use, fluorescein diacetate (Sigma Chemical Co. or other supplier) is dissolved in acetone, 10 mg in 5 ml. This solution is added drop by drop to a few ml of 10% sucrose in a small vial until the first permanent milkiness is seen. Only two or three drops are usually required. The sucrose concentration must be adjusted to suit the tonicity of the pollen, for example grass pollen will require 30% sucrose.

For use dissect the pollen straight from the anther into the medium. Leave for a moment, then cover with a coverglass, and observe by fluorescence microscopy. Observations must be made within about 10 min, otherwise fluorescence fades as the pollen grains lose viability in the medium. The technique thus depends on the integrity of the plasma membrane and was developed for pollen by J. and Y. Heslop-Harrison in 1970.

7.3 Demonstration of pollen-wall proteins

Proteins diffusing from pollen include the allergens. They can be demonstrated in two simple ways. First: *pollen prints* can be made on agar films on microscope slides. Prepare 1% agar, and place 3 drops of melted agar on a slide with a pasteur pipette, allowing it to spread into a flat film. Dust dry pollen onto the adhesive side of sellotape, and press gently into contact with the agar film. Leave in contact for a few seconds or for up to 15 min. Observe after flooding film with Coomassie Blue stain-fixing medium and covering with coverglass. By light microscopy, sites of protein release into film will be seen as blue areas. These will often mirror the distribution of germinal apertures in the pollen.

Secondly, the actual release can be observed under the microscope by flooding the grains with the stain-fixing medium. Place a piece of double-sided sellotape on a slide, dust on some pollen (*Cosmos*, sunflower or other Compositae are especially interesting) and cover with cover glass. Observe the dry pollen towards the edge of the coverglass, while pipetting some stain-fixing medium under the coverglass. The grains will be seen to swell, followed by immediate release of their wall proteins, first from exine sites all around the grains, then after a few minutes, from the intine at the germinal apertures.

Stain-fixing medium: prepare 0.25% solution of Coomassie Blue in 45% methanol and 10% acetic acid. When dissolved, dilute 1 part with 3 parts of 20% sucrose and filter immediately before use.

7.4 Demonstration of pollen grains in honey

Weigh out 10 g honey and dissolve it in 20 ml of water warmed to 30–40°C. Place in centrifuge tube, and centrifuge at low speed for 10 min. Draw off most of the supernatant liquid, leaving 1–2 cm above the sediment. With a Pasteur pipette, place a drop of the sediment on a microscope slide, add Calberla's fluid, mix and apply coverglass. Pollen present will be stained a red colour, and can be identified and even scored to show frequency of particular types in the honey.

7.5 Pollen germination and demonstration of callose

Many pollen grains readily germinate in simple media and tubes can be observed by light microscopy without staining. Prepare 1% agar containing 10% sucrose and 0.008% boric acid, pipetting a few drops onto a slide to make a flat film. When cool, sprinkle with pollen and store in moist Petri dish for germination to occur. Pollen of lily, fuchsia and *Impatiens* germinates readily.

To demonstrate the presence of the polysaccharide callose in germinating pollen and pollen tubes, simply stain the tissue in decolorized aniline blue, and observe by fluorescence microscopy using a blue or UV exciter filter. The callose will fluoresce a bright yellow or blue colour depending on the exciter filter. It is always essential to check staining with tissue mounted only in water in case of autofluorescence. The stain is prepared by adding 2.3 g of tripotassium orthophosphate (i.e. 0.1 M K_3PO_4) to 100 ml of distilled water, followed by 0.1 g water soluble aniline blue. The bottle is stoppered and placed in a dark place to decolorize for 24 h. For use it should be clear, and glycerol can be added to 10% by volume to prevent the stain drying out. The mounts are prepared and stored in the stain.

Further Reading

ERDTMAN, G. (1968). *Handbook of Palynology*. Hafner Press, New York.
ERDTMAN, G. (1971). *Pollen Morphology and Plant Taxonomy in Angiosperms*. Hafner Press, New York.
FAEGRI, K. and VAN DER PIJL, L. (1971). *The Principles of Pollination Ecology*. Second edition, Pergamon Press, Oxford and New York.
FAEGRI, K. and IVERSEN, J. (1975). *Textbook of Pollen Analysis*. Third edition, Hafner Press, New York.
GODWIN, SIR HARRY. (1975). *The History of the British Flora. A factual basis for phytogeography*. Second edition, Cambridge University Press, London.
GREGORY, P. H. (1973). *Microbiology of the Atmosphere*. Second edition, Leonard Hill, Bucks., England.
GUNNING, B. E. S. and STEER, M. (1976). *Plant Cell Biology – an ultrastructural approach*. Edward Arnold, London.
HESLOP-HARRISON, J. (ed.) (1971). *Pollen, Development and Physiology*. Butterworths, London.
HESLOP-HARRISON, J. (1972). Sexuality of angiosperms. In: *Plant Physiology, a Treatise*. Ed. F. C. Steward. Vol. 6C. Academic Press, London and New York.
HESLOP-HARRISON, J. (1975). Incompatibility and the pollen-stigma interaction. *Ann. Rev. Plant Physiol.*, 26, 403–25.
HESLOP-HARRISON, J. (1978). *Cellular Recognition Systems in Plants*. Studies in Biology, no. 100, Edward Arnold, London.
HODGES, D. (1974). *The Pollen Loads of the Honey-bee*. Bee Research Association, London.
HYDE, H. A. and ADAMS, K. F. (1958). *An Atlas of Airborne Pollen Grains*. Macmillan, London.
KEMP, R. (1970). *Cell Division and Heredity*. Studies in Biology, no. 21, Edward Arnold, London.
KNOX, R. B. (1976). Cell recognition and pattern formation in plants. In: *The Developmental Biology of Plants and Animals*. Eds. C. F. Graham and P. F. Wareing, pp. 141–67. Blackwell, Oxford.
MORCOMBE, M. K. (1968). *Australia's Western Wildflowers*. Landfall Press, Perth.
NILSSON, S. (ed.) (1973). *Scandinavian Aerobiology Bull. 18*, Swedish Natural Science Research Council, Stockholm.
OGDEN, E. C., RAYNOR, G. S., HAYES, J. V., LEWIS, D. M. and HAINES, J. H. (1974). *Manual for Sampling Airborne Pollen*. Hafner Press, New York; Collier Macmillan, London.
STANLEY, R. G. and LINŞKENS, H. F. (1974). *Pollen, Biology, Biochemistry, Management*. Springer-Verlag, Berlin, Heidelberg, New York.
STRAKA, H. (1975). *Pollen-und Sporenkinde*. Gustav Fischer Verlag, Stuttgart.
WODEHOUSE, R. P. (1935). *Pollen Grains*. Reprinted by Hafner Press, New York.